The Digital Horizon: Hopeful or Hopeless

HOW AI TECHNOLOGY AND HUMANITY ARE EVOLVING TOGETHER

ALEX N. BEAVERS, JR., PhD

Palma Sola Publishing

This book contains information obtained from authentic and highly regarded sources. Reprinted materials is quoted with permission, and sources are indicated. A wide variety of references are listed. Reasonable efforts have been made to publish reliable data and information, but the author and publisher cannot assume responsibility for the validity of all materials or for the consequences of their use.

Trademark Notice: Product or corporate names may be trademarks or registered trademarks, and are used only for identification and explanation, without intent to infringe.

Copyright © 2017 by Alex N. Beavers, Jr.

All rights reserved

Published in the United States by Palma Sola Publishing

Library of Congress Cataloging-in-Publication Data

Beavers, Jr., Alex N.
  The Digital Horizon: Hopeful or Hopeless / Alex N. Beavers, Jr.
    p. cm.
  Includes bibliographical references and index
  ISBN-13: 978-1544725451
  ISBN-10:1544725450

Dedication

To Chris who helped me understand lifelong adaptability. To Alisa who helped me understand how every person has a story to discover. To Amanda who helped me understand how passion and patience work together in our digital world. To Robyn who helped me think beyond the visible horizon. To Kal who helped me understand how innovation can be brought to the education industry. To Brian who helped me understand how digital technology can be a helpful part of our daily lives. To Liam who helped me understand how a few well-chosen words are more effective than a deluge of them. To David who helped me understand that the human mind has its own maturation schedule.

And most of all to Linda, the bright sun on my horizon.

Table of Contents

List of Exhibits vii
Preface ix
INTRODUCTION 1
1. Gloom vs Bloom 3
    Technology Teeter-Totter 3
    The History of Gloom 6
    The History of Bloom 9
2. What is the Digital Horizon? 13
    Fundamental Technical Elements 13
    Computation 15
    Networking 16
    Actuation 17
    Sensors 19
    Data 21
    Algorithms 23
    Memory 24
    Power 26
    Systems and Synergy 27
ECONOMICS: If No ROI, Then No AI 31
3. The Productivity Cycle 33
    Humanity's Drive to Be Better 33
    Standard of Living 36
    Investment 37
    Technology Deployment 40
    Productivity Improvement 42
    Wealth Creation 44
    Law of Diminishing Returns 46
4. The Innovation Cycle 51
    Need and Opportunity 51
    Innovation 52
    Commercialization 52
    Economic Transformation 54
5. Cycle vs Cycle 57

  Living Standard Stimulates New Needs 57
  Investment in R&D and CAPEX Fuels Innovation 58
  Commercialization Needs Deployment Success 59
  Economic Transformation Fuels Wealth Creation 60
6. How Does Technology Improve Productivity? 61
  Technology Alone Is Seldom The Solution 61
  Designing The Workflow Is Essential 64
  Eliminating Variation Is Essential 68
  Including Flexibility Is A Virtue 70
  Big Data Is Not Necessarily Big Information 72
  Decentralization Is The Future 77
INTELLIGENCE: Masters Of AI By Design 81
7. What Does It Mean to Be a Machine? 83
  Do Machines Think or Perform? 83
  A Few Thought Experiments 86
  Games 86
  Medical Diagnosis 88
  Investment Decisions 89
  Scientific Discovery 91
  Machine Self-Preservation 91
  Was Sherlock Holmes the First AI Machine? 92
8. What Does It Mean to be a Human? 95
  Evolution Of Human Intelligence 95
  Human As Master Of Machine 98
9. More Responsibilities for Machines? 101
NECESSITY: Tech Drives Standard Of Living 109
10. Human Need Demands More Technology 111
11. Industries That Need Major Improvements 115
  Agriculture - A Major Success Story 115
  Education - In Need Of Success 118
  Healthcare - Improvements Urgently Needed 121
  Government - Always in Need of Improvement 126
  Energy - Improvements are Happening 130
  Water - Many Drops to Drink 135
  Transportation - Disruption Ready 139
  Other Service Industries 144

EVOLVING TOGETHER	149
12. Machine Evolution	151
13. Human Evolution	155
IMPORTANT QUESTIONS AND ANSWERS	159
14. How Far Off Is The Horizon?	161
Technology Adoption Rates	161
Thought Experiment: Hamburger Automation	164
15. Does the Economic System Matter?	169
The Short Answer: Yes.	169
16. How Fragile is the Digital Horizon?	177
Cyber Security Risks	177
Physical Risk	183
Cosmic Energy Risks	184
Economic Risks	186
A Outline For Reducing Digital Fragility	187
17. Will People Be Left Behind by Automation?	189
Because Of Income Inequality?	189
Because Of Job Destruction?	194
Because Of Education?	197
Who Does Get Left Behind?	199
18. Is Innovation Needed in Social Services?	201
The Short Answer: Yes!	201
Is Basic Income A Solution?	205
19. Will Automation solve all problems?	209
The Short Answer: No!	209
Despite More Abundance?	209
Because Of More Decisions?	210
Even If It Did, Why Get Up In The Morning?	210
CONCLUSIONS	213
20. Digital Horizon Is Hopeful	215
References	221
About the Author	229
Index	231

List of Exhibits

Exhibit 1.1 Technology Gloom vs Bloom 3
Exhibit 2.1 Binary Representation of the Real World 14
Exhibit 2.2 The Digital Horizon Technology Elements 15
Exhibit 2.3 Computation Technology Advancements 16
Exhibit 2.4 Networking Evolution 17
Exhibit 2.5 Actuation Evolution 19
Exhibit 2.6 Sensor Evolution 21
Exhibit 2.7 Growth of Data Moved and Stored 23
Exhibit 2.8 Digital Memory Trends 25
Exhibit 2.9 Battery Trends 26
Exhibit 2.10 Printed Circuit Technology Evolution 28
Exhibit 3.1 Examples of Humanity's Improvements 33
Exhibit 3.2: Maslow's Hierarchy of Needs 34
Exhibit 3.3 The Productivity Improvement Cycle 36
Exhibit 3.4 US Value Creation Investment Types 38
Exhibit 3.5 Sources of R&D Spending 39
Exhibit 3.6 Advertising Industry Transformation 43
Exhibit 3.7 Simple Farm Example of Diminishing Returns 46
Exhibit 3.8 Productivity Improvement and Diminishing Returns ... 47
Exhibit 3.9 US Labor Productivity History 48
Exhibit 3.10 Innovation Overcomes Diminishing Returns 49
Exhibit 4.1 The Innovation Cycle 51
Exhibit 4.2 Music Industry Sales in Units 55
Exhibit 5.1 Innovation and Productivity Cycle Relationships 57
Exhibit 6.1 Manual Torque Wrench Distribution Plot 70
Exhibit 6.2 Centralized vs Decentralized Structures 79
Exhibit 10.1 Population Aging Factors 111
Exhibit 11.1 Grain Output Productivity Growth 115
Exhibit 11.2 Milk Output Productivity Growth 116
Exhibit 11.3 US Farmer Productivity Growth 117
Exhibit 11.4 Pre-College Education Industry 118
Exhibit 11.5 Post-Secondary Education Industry 119
Exhibit 11.6 Healthcare Industry Spending 122

Exhibit 11.7 Healthcare Spend by Service Type 124
Exhibit 11.8 US Federal Government Staffing 127
Exhibit 11.9 Employee Work Time Comparison 128
Exhibit 11.10 US Energy Productivity Trends............................. 131
Exhibit 11.11 Energy Productivity Drivers................................... 131
Exhibit 11.12 US Energy By Source and Sector 132
Exhibit 11.13 Water Uses in the US 136
Exhibit 11.14 Consumer Water Uses (Indoor) 137
Exhibit 11.15 Vehicles per Capita.. 139
Exhibit 11.16 Average Load Factors... 140
Exhibit 11.17 Fares vs Operating Costs..................................... 141
Exhibit 11.18 Full Autonomous Fleet Impact 142
Exhibit 11.19 Services vs Manufacturing Productivity.............. 144
Exhibit 11.20 Industry Segment Productivity 145
Exhibit 12.1 Machine Evolution Model....................................... 152
Exhibit 13.1 Evolution Of Human Interaction With Machines...... 155
Exhibit 13.2 Symbiotic Machineand Human Evolution?............ 158
Exhibit 14.1 Technology Adoption Rates 162
Exhibit 14.2 Hamburger Workflow Activities List 165
Exhibit 14.3 Hamburger Business Processes 165
Exhibit 14.4 Manual vs Automation Cost Models....................... 166
Exhibit 15.1 Competitiveness Ranking of Top Economies............ 171
Exhibit 15.2 Global Economic System Model 172
Exhibit 15.3 Reshoring Trend 175
Exhibit 16.1 SEU Failure Rate Trends 185
Exhibit 17.1 Income Inequality 190
Exhibit 17.2 Average Income Comparisons 191
Exhibit 17.3 Upward Income Mobility of US Taxpayers 192
Exhibit 17.4 Occupations at High Income Level......................... 193
Exhibit 17.5 Job Creation/Destruction/Mobility Model 194
Exhibit 17.6 140 Years of UK Job Data 195
Exhibit 18.1 Social Spend as % of OECD GDP............................ 202
Exhibit 20.1 Digital Horizon Meter Points To Hopeful................ 216

Preface

Technology is the stuff of delightful dreams and scary nightmares. Steam power brought the industrial revolution, simultaneously raising living standards but also upending societies. Powered flight gave us both the joy of traveling the world and the terror of London's Blitz. More recently, the internet has freed communication more than ever while leaving us vulnerable to all manner of new threats prefixed with "cyber-".

Today, we grapple with a similar conundrum: Is humanity's microprocessor-based, big data driven, artificial intelligence (AI) intensive future -- the "digital horizon" -- full of hope or devoid of it?

The dream is more likely than the nightmare. Our digital horizon is more likely hopeful rather than despairingly hopeless. There are three reasons for the hopefulness.

The first reason is **economics**. History has shown that for a new technology to enter society and be sustainable, it must have a beneficial economic impact. If a new technology improves productivity, then new wealth is created and quality of life improves. In other words, if there is no return on investment, then no new technology. (If no ROI, then no AI!)

The second reason is **human intelligence** (HI). The human version of intelligence is building the artificial version in its own image using tools such as algorithms, neural network structures, and data analytics and teaching it with data from sensors and stored human knowledge. These tools are just mathematical approximations of what we think makes the universe tick. We can and are building machines that execute these tools many times faster than the human brain. The technology we are building goes under a variety of names such as machine intelligence, robotics, or automation, but the scariest name used is artificial intelligence. No matter which name is used, the artificial version of intelligence is faced with the same challenges as the human version: having to make decisions with imperfect data collected from an uncertain world. We can and will make the tools that we design into artificial intelligence better, but they will still be our tools. In other words, since we are designing the new technology, it will not on its own accord become our alien AI overlords. (HI masters AI by design!)

The third reason is demographic ***necessity***. To offset and diminish the chances of a severely diminished quality of life due to an aging population, there is a need for more, not less, new technology to sustain and accelerate productivity growth. In other words, if there is not a continuing stream of new technology to continuously improve our economic productivity, then our standard of living (SOL) will decline. (SOL needs AI)

These three reasons should prevent humanity from being overmastered by its own creation. History has shown us that there is a symbiotic relationship between technology and society. Technology is changing how we behave, learn, and adapt biologically while we make technology work to further our needs and desires. All indications are that mankind and technology will evolve simultaneously and synergistically as we sail toward the digital horizon. (AI and HI are evolving together)

SECTION I

INTRODUCTION

Chapter 1

GLOOM VS BLOOM

Technology Teeter-Totter

The advancement of civilization is almost always measured by milestones of technological innovation, starting with the first stone tools. It is not clear how many genes might be associated with curiosity, inspiration and creativity, but history indicates that there were moments when a few humans with exceptional mental gifts (and often exceptional luck) would make a breakthrough in knowledge and/or knowhow propelling homo sapiens to the next level of intellectual and physical wellbeing.

Despite this history of using technology innovation to mark a sociological step upward, there is always a downward force of opinion and fear that creates gloom about eventual outcomes (Exhibit 1.1).

EXHIBIT 1.1 TECHNOLOGY GLOOM VS BLOOM

Even as we hope for more marvels on the horizon, we fear the side effects they may bring.

The wirelessly-connected socially-networked data-streamed world in which we now live has nurtured our anxiety. The historic fear of the scientific unknown that has been the subject of works of science fiction foretelling invasions of alien species, meteorite collisions and nuclear holocausts is now being supplemented by Internet headlines of pandemics, biblical weather events and geopolitical turmoil.

One gloomy outlook which has picked up momentum is the fear of un-throttled technology achieving domination over humanity. Whether in the form of Terminator-like robots or in the form of artificially intelligent software that manipulates cloud-stored data through ubiquitous communications, a growing list of celebrities from government and industry have voiced concerns about the potential for a mechatronic apocalypse. The fears range from automation replacing all tasks now performed by humans to an artificial intelligence that decides on its own to systematically eliminate what it considers a superfluous human species.

Forecasts and alerts about a digital horizon where human endeavor would be replaced by unblinking machine efficiency have become more frequent over the last few years. Whether couched in an economic context (all jobs being eliminated) or in a philosophical context (Is mankind necessary?), these doomsday warnings seldom have much thought put behind them beyond what an addled (or richly developed) imagination can supply.

What makes these alerts worthy of consideration is that some are coming from very successful innovators and entrepreneurs such as Bill Joy (Joy, 2000) (founder and original CTO of Sun Microsystems), Elon Musk (Price, 2015) (Co-founder of PayPal, Tesla, and SpaceX), and Steve Wozniak (Smith, 2015) (Co-founder of Apple Computer).

Also, there has been a steady stream of achievements by private corporations publicized with titles that imply an impending takeover by intelligent machines. IBM, for example, has a machine intelligence it has named "Watson" in honor of the early and very successful leader of the company, Thomas J. Watson. The machine has become famous for winning against the two best Jeopardy TV show contestants (Markoff, Computer Wins on 'Jeopardy!': Trivial, It's Not,

2011). More worrisome is that Watson was the technical "offspring" of IBM's "Big Blue" computer that won a chess tournament against grandmaster Gary Kasparov in 1997 (Finley, 2012).

On the other hand, the new generation of high tech giants such as Google, Facebook, Apple, and Microsoft have been investing heavily in the form of seed investments and acquisitions of robot and artificial intelligence (AI) companies as well as in internal research (Forrest, 2014) and (Albergotti, 2014). Even the world's largest intellectual cocktail party, the World Economic Forum in Davos, in 2016 featured several discussions on why the world should love intelligent machines and not fear them (Gilbert, 2016). Of course, many of the speakers voicing this theme at the conference were representatives of the tech companies developing and selling AI products.

Then there are the optimists that tend to the extreme in their outlook about technology. Their views often wander into the world of the unicorn, foreseeing a magical future where all problems are solved by conjectured breakthroughs. One of the more outspoken technical optimists, Peter Diamondis (founder of Singularity University), has for several years envisioned a future of abundant goods, services and wealth brought about by an exponential expansion of technology in all realms which solves all pressing current and future problems (Peter Diamondis, 2012). While unbridled optimism is a welcome relief from the gloomiest clouds of the doomsayers, it does run the risk of overlooking some of the big questions that will likely emerge from the continued evolution of technology.

This book posits a third outcome. In full disclosure, my original hypothesis centered on the optimistic side of the debate. After working through the data, studying historical lessons learned, and discussions with thoughtful people from a variety of disciplines and backgrounds, it became clear that while the digital horizon beckons something short of an unbridled optimism, it does represent a hopeful future that should not be feared and which must be pursued. The troublesome questions that await us in the future are not all that different from those that we have asked in the past. Digital technology and the innovative souls that create it in all its forms offers a bright blooming future that has answers to many of our old questions and tools for solving the new ones.

The History of Gloom

People voicing fear about the impact of machinery and automation on the elimination of jobs and the dehumanization of the workforce goes back to before the Industrial Revolution and began about the same time as the first machines were deployed in economic activities (Lewis J. , 2016) and (Morgenstern, 2016). Early on, fear of automation -- or industrialization as it was called at first -- was provoked by the assumption that the demand for labor was inelastic. Even worse, the assumption was that there were only so many jobs in existence and if one job was eliminated due to automation, there would be no new jobs created elsewhere to offset that loss. The job market was assumed to be a finite-sum game.

In more recent times, the concept of elasticity of demand and supply of goods, services, and labor provided for more optimistic perspectives on job disruption and creation. Yet this optimism has been offset in the minds of many thought leaders by the concern that the acceleration of the use of technology to replace old jobs was outpacing the creation of new ones.

In the late 1500's, Queen Elizabeth I of England would not approve a patent for an invention by William Lee of the first stocking frame knitting machine even though it improved the productivity of hand knitters by a factor of more than 20 times. The queen reportedly was concerned that the invention would throw too many of her subjects out of work and turn them into beggars (Acemoglu & Robinson, 2012). Lee left England to set up his operations in France. Eventually, Lee's invention made its way back to England where its technology became the backbone for the wool, silk, and lace knitting industry – but only after a century had passed.

In the early 1800's, British economist David Ricardo believed that the cost of production primarily set the price of goods rather than supply and demand (Ricardo, 1812). Thus, if a hat maker could suddenly increase his hat production rate by a factor of 2, the price of each hat would diminish by 50%, thus keeping his hourly income at the same level or driving it down it down if he could not sell the extra hats. Ricardo voiced his concern that productivity improvement from technology was likely to drive wages down.

Also in the early 1800's, the Luddite movement among British textile workers emerged. This activist, sometimes militant, group fought against worsening working conditions and threats to their jobs. They believed this was brought on by the continued and ruthless application of technology to improve the productivity of their industry (UK National Archives, 2016). Although the use of human-powered mechanical looms had been solidly established for decades, it was the application of the newly invented steam engine, resulting in a sudden explosion in productivity, that created a major fear amongst loom workers. Today, of course, to be a Luddite is to be opposed to automation in general.

In the mid-1800's, Karl Marx gloomily wrote about the polarization of capitalism being driven by the application of labor replacing machines. He was one of the first writers to describe how industrialization could change the social structure by defining two social classes: the bourgeois who owned industry and an exploited working class that labored in it. He outlined how these two classes would be at odds over the creation and sharing of wealth and should eventually be in conflict. He did not foresee or appreciate how an industrialized capitalistic system would evolve that did share wealth and did create abundance of goods and services on a massive scale.

In the 1930's, the English economist John Maynard Keynes described the short-term effect of job dislocation from the application of technology as "technological unemployment". In the midst of the Great Depression of the 1930's, he was an optimist about the likely positive economic impact for the future of industrialization. However, he did voice a concern that there would be a general pessimism due to a "temporary phase of maladjustment" as society and the economy adjusted to ever greater levels of productivity (Keynes, 1963).

In 1964, a group of notable scientists and economists known as the Ad Hoc Committee on the Triple Revolution sent President Lyndon Johnson a letter forecasting the danger of a revolution triggered by automation that could divide society into a skilled elite and an unskilled underclass (Revolution, 2016).

It is not only technology that has created a pessimism about job elimination. Jobs can be eliminated in one geographic area when they are replaced by automation or when they are moved to a different geographic area that has lower wage rates. The disruption that occurs when a formerly employed person must find new employment and/or get new training and education is personal, emotional, and economic in nature. While the business decisions that are made by corporate executives might be rational in a global economic sense, the impact on a local economy can appear irrational. Technology is responsible in many ways for the growth of global trade by making it easier to relocate plants, to build new plants with new automation, to transmit money and information electronically, and to transport large quantities of materials and products efficiently. By making global trade more practical and productive, technology can make competitive factors out of supply chain structures, job content, automation, materials cost, labor cost, and government policy. While "off-shoring" is often viewed as primarily a political issue (as illustrated in the surprise results of the 2016 US presidential election), technology does have a role in the global economic trends that create disruptions in job content and location. Thus, pessimism about technology can be exacerbated by global trade trends.

To date history has proven that technology does eliminate jobs in the short term and in specific industries, but in a longer term also creates many more jobs in new categories in related or unrelated industries. Productivity improvement from technology does lead to new wealth creation and improved standards of living in general and on the average (Mokyr, Vickers, & Ziebarth, 2015).

The History of Bloom

By most statistical measures, each new addition of automation in the economy has benefited every economic level. Looking at economic data from Britain, where the battle between job creation and industrialization has the longest history, real incomes approximately doubled during the millennium prior to 1570. They then tripled in the 300 years from 1570 to 1875, and then more than tripled in the 100 years from 1875 to 1975 (Kambayashi, 2014). Even Keynes in his 1930 essay about his vision of the future described technological unemployment as a short-term issue, predicting that in the longer term, everyone becomes richer.

In general, the number of jobs has increased with the introduction of new plant, equipment, and other technology. Average wealth per capita (income, etc.) has risen with the introduction of new automation. The average lifespan has increased as more technology reduces physical risks and improves healthcare. The average number of hours worked per week has decreased. Even average IQ has increased as everyday life benefits from new technology (more about this later).

It is not easy to quantify optimism versus pessimism in literature or the media especially when it comes to topics related to technology. Pessimists are often viewed as fearful worriers while optimists are often viewed as fanciful dreamers. A slogan sometimes used by financial advisors and self-help experts, "doom and gloom sells, optimism pays", might describe our fascination with messages of fear while we work for the future. What we do as individuals and as a society might be driven by instincts of self-preservation, delusions of grandeur, or random events. The common denominator is the desire to do better.

Doing better can mean making people's lives better with more food, more safety, more healthcare, more entertainment, or more recreation. Doing better can mean making a company more profitable and making investors and employees wealthier. Doing better can mean making more tax revenue for governments, more tuition revenue for schools, more healthcare revenue for hospitals. Doing better can mean making armies more powerful, making peacekeepers more effective, making doctors more effective. Doing better can mean making the life of children better than it was for their parents.

One economic metric can be used to measure how much better we are doing: productivity. Whether it be the productivity of labor, machines or money, it is productivity's inherent promise -- doing more with less -- that is the key element in the human decision processes leading to the digital horizon.

The digital horizon is formed by the technologies that are based on the digital sciences: the computer chips that process and handle data; the electrical, mechanical, and chemical machinery that they control; the information they capture, store, and communicate; and the decisions they are programmed to make. It includes mechatronic automation, artificial intelligence, the Internet of Things, and Big Data Analysis.

In many ways, the developments taking place in the life sciences, materials sciences, and physical sciences are colliding with and being enhanced by digital technology. Few scientific discoveries and product/service developments could be achieved today without automated equipment, robotic devices, data sensors, analysis algorithms, and computer models. Thus, whatever we find at the digital horizon will include the digitally stored, analyzed, and implemented results from the expanding fields of scientific, healthcare, and information research.

The advancement of the life sciences, such as genetic engineering, bioinformatics, cloning, epigenetics, etc., and the social sciences often call into question, either explicitly or implicitly, what it means to be human (Bjorkland, 2007). When considering the digital horizon, a similar and related question emerges: what it means to be a machine.

These two questions and their answers are ultimately interconnected and cannot be addressed independently. The analysis presented in the following chapters is focused on the technical and economic processes behind the formation of the digital horizon. The key social questions that are likely to emerge are also highlighted and examined.

The objective herein is to examine the types of decisions and practical motivations that are creating the pathway to the digital horizon. The goal is to provide a platform for reasoned discussion about how technology has been deployed in the past and may well be in the future. From this vantage, it is possible to think about the very important questions facing society rather than reading and filing away the latest doomsday (or unicorn optimistic) headlines.

Chapter 2

WHAT IS THE DIGITAL HORIZON?

Fundamental Technical Elements

While all the technologies discussed in the following chapters are inherently digital, the universe and all its creatures are definitely not. Since necessity is the mother of invention, evolution has led us down a path of creating an expanding portfolio of sciences and mathematics which that have helped us learn about, model, simulate, control, and exploit the physical world.

The first inventions that used electricity for practical purposes were made over 100 years ago, but an appreciation of the full potential of electronics did not emerge until World War II. The earliest forms of electronic technology were analog in nature: electrical signals were continuous waveforms. Eventually, digital electronic technology became dominant because the productivity of silicon based semiconductors in terms of performance/price ratio grew exponentially.

The now famous "Moore's Law" that was coined by Gordon Moore, one of the founders of the semiconductor industry, in 1965 (Moore, 1965) to describe the rapid growth in the power of semiconductors says it all: "The complexity for minimum component costs has increased at a rate of roughly a factor of two per year". Moore predicted that this doubling of performance for the same cost every year would last for at least 10 years. This forecast of a doubling in price/performance has endured for the last 5 decades although the time factor has changed a bit (it is more like a doubling every 3 years now).

The Digital Horizon: Hopeful or Hopeless

Digital technology requires that all information be converted to or approximated by binary numbers, i.e. 1's and 0's. Binary representation of information became necessary because the calculators (originally mechanical but eventually electronic) that were invented to speed up the recording and calculation of numbers were based on the Boolean logic of switches that were either on or off. Once it was realized that binary numbers were the best way to connect the strengths of solid state electronics and computational algorithms, the exploitation of mathematics in engineering sciences accelerated.

As physical and social scientists continue the search for truth about the universe, engineers are busy designing and making everything as digital approximations of the known reality (Exhibit 2.1). Binary representation of the physical real world is thusly the cornerstone of the technologies that are promising to be hopeful or threatening to be hopeless.

EXHIBIT 2.1 BINARY REPRESENTATION OF THE REAL WORLD

Heron on Roof Top

16 Color (4 bit) Digitized Image of Heron on Roof Top

Binary Data Form of Digital Image of Heron on Roof Top

Throughout this book, a variety of terms such as artificial intelligence, machine learning, robotics, and automation are used interchangeably as names for the technical forces taking us to the digital horizon. This is done for convenience since they all represent different perspectives of the same thing: a smart machine.

Artificial intelligence tends to make us think of a thinking machine. A robot tends to make us think of a mobile and agile android machine. Automation tends to make us think of a programmable machine tool tending a conveyor belt bolted to the floor of some far-off factory. At the heart of each of these is a core set of technologies (Exhibit 2.2) that are continue to become less expensive, smaller, more powerful, and therefore, more productive.

EXHIBIT 2.2 THE DIGITAL HORIZON TECHNOLOGY ELEMENTS

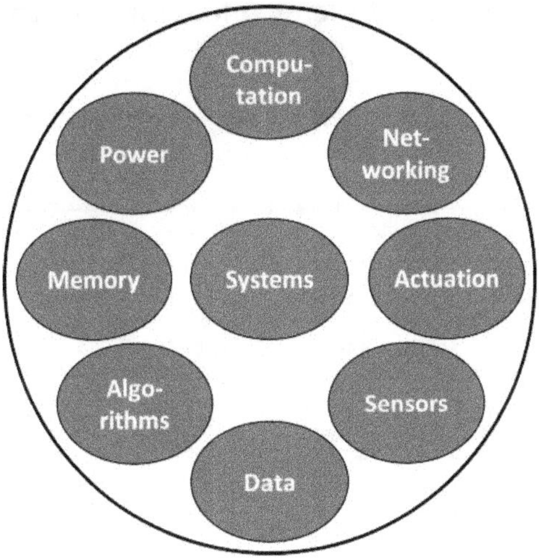

[Diagram: A circle containing nine labeled ovals arranged with "Systems" in the center surrounded by "Computation", "Networking", "Actuation", "Sensors", "Data", "Algorithms", "Memory", and "Power".]

Computation

Probably the most important technical element in the digital horizon is the decreasing cost and size and increasing power of digital electronic circuitry. Decades ago, silicon-based integrated circuits (or ICs as they became fondly known) replaced vacuum tubes as the active digital components at the heart of every supercomputer, mainframe computer, minicomputer, personal computer, smart phone, tablet, and every other smart device.

In 1980, microprocessor ICs had an operating speed of a few thousand instructions per second and were enclosed in a plastic case that had a footprint of several square centimeters and cost on the order of $100 or more each. Today, a microprocessor runs millions of times faster, occupies a space of a few square micrometers, and has a cost measured in pennies rather than $100 bills.

Computational circuits are small enough, cheap enough, and consume such low amounts of power that they can be integrated into almost device that can be engineered: aircraft, automobiles, home appliances, cell phones, thermostats, electrical outlets, medicine tablets, and sensors.

Now almost anything can be made "smart" because of the productivity achieved by the semiconductor industry over the last 40 years (Exhibit 2.3).

EXHIBIT 2.3 COMPUTATION TECHNOLOGY ADVANCEMENTS

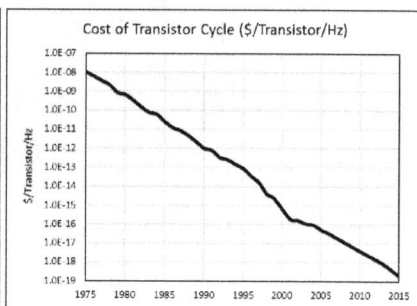

Sources: IBM, Intel, HP, Dataquest, Singularity

Networking

Although the Internet is the result of the evolution of networking technology over decades, the usefulness of the networking of electronic devices has become obvious to every man, woman, and child on the planet. This usefulness continues to grow as each new advancement in cost effective networking technology is deployed.

When networking technology broke free of the constraints of wires, entire countries could leapfrog generations of grids of wires strung from wooden poles. Wireless networking technology, whether it be in the form of cellular, Wi-Fi, Blue Tooth, or other protocols, can be deployed globally in a relatively short period. Now not only can every type and size of device be made smart, they can also be wirelessly networked.

Networked and intelligent sensors, materials, products, and systems have created the evolving digital world that has things such as home automation, self-driving cars, smart highways, distance learning, and smart medicine that we expect to include in our future. This extended reach of machine intelligence provided by sensors has given birth to the term "Internet of Things" or IoT (Exhibit 2.4).

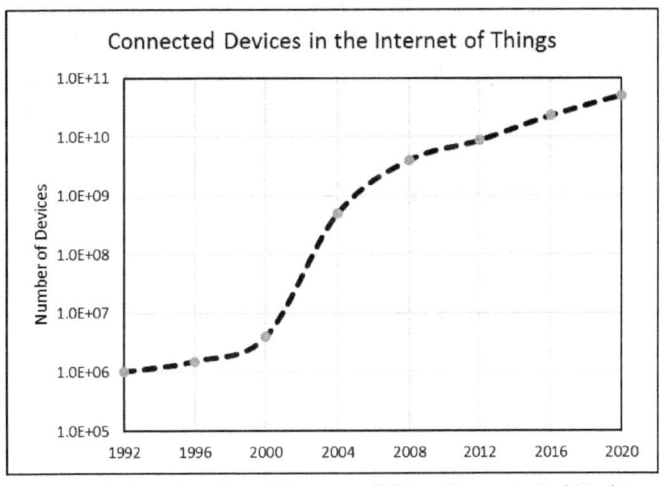

EXHIBIT 2.4 NETWORKING EVOLUTION

Source: Forbes, Roundup Of Internet of Things Forecasts And Market Estimates, 2015

This is already occurring and the number of networked devices is expanding rapidly. A typical upscale home today that houses 4 people (say two parents and two children) might have 3 or 4 laptops, one or more streaming TV devices, one or more wireless audio speakers, one or more wireless video game, one or more wireless thermostats, one or more wireless kitchen appliances, one or more cars with wireless features, and all connected via a wireless router. In just one home a wireless network of 15 to 20 "things" that can send and receive information from the internet.

Actuation

While computation and networking technologies make devices smarter and more collaborative, actuation technology makes them move and perform physical tasks. Actuation is the "muscle" in the world of automation and artificial intelligence.

Actuation technology includes electro-mechanical technologies such as electric motors, magnetic valves, hydraulic motors and pistons, and electroactive materials. The electro-mechanical actuation industry has continued to improve in its price/performance ratio as well as in breadth of application. Three decades ago, the robots were large. They were built to handle large metal parts weighing thousands of pounds for automobile production. To provide this amount of power at the time, the well-known technology of hydraulic systems was used. Since that time, electric motors have replaced hydraulics. The electric motors allowed smaller, more cost effective, and more flexible robotic systems to be built. The cost of a giant robot "arm" 30 years ago would have been in the hundreds of thousands of dollars. In 2016, there are several venture-backed companies working on robot arms for consumer applications that would cost in the range of a hundred dollars.

Very small electric motors are used to actuate disk drive trays, eyelids in dolls, or joints in robot hands. These small motors can be designed for simple applications (e.g. toys, disk drive trays) and cost only pennies; or they can be designed for complex applications that require high precision (e.g. surgical robot fingers) and cost hundreds of dollars.

Emerging new technologies for actuation include new materials such as electroactive polymers. Electroactive polymers (EAP) are polymers (often very elastic plastics) that change in size and shape when placed within an electric field such as when an electrostatic charge is placed on its surface. It is much like a capacitor (typically a small electronic component that stores electric charge and is used within electronic circuitry) that changes shape. It was well known for decades by electronic component designers that electrostatic fields create mechanical forces within capacitors that could make it change shape as a function of the applied voltage. Since having shape-changing electronic parts within a television or cell phone was not generally a desirable design feature, the goal of engineers for decades was to design the parts that were as rigid as possible. It was not until scientists started thinking about how to harness these forces rather than suppress them that the application of electroactive polymers became apparent. These types of actuators are just beginning to appear as actuators in very small format devices such as robot fingers and vibrating game controllers (Exhibit 2.5).

EXHIBIT 2.5 ACTUATION EVOLUTION

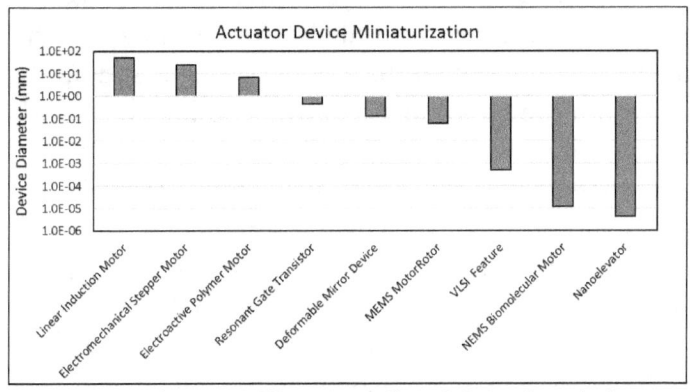

Sources: Singularity.com, IEEE Transactions on Electronic Devices, Nanotechnology

Sensors

The same improvements in semiconductor technology that have made computational devices so powerful and inexpensive have made it possible to design sensor chips that can measure temperature, pressure, humidity, smoke, toxic gases, physical movement, and physical surfaces. Camera, audio, laser, and ultrasonic emitter chips are now the size of dust particles and can be incorporated into any device, vehicle, or structure.

Electronic sensors digitize physical phenomena (i.e. generate data) and communicate it to computational devices within a system or somewhere across a network. Since a broad range of physical phenomena can be digitized, sensors can give machine intelligence the five human senses of sight, hearing, smell, touch, and taste. By giving machines the five senses, the impact that computer software can have over the physical world grows exponentially.

One example of the impact is in the heating and air conditioning of homes or offices. The advent of wirelessly networked thermostats was spawned by the Silicon Valley company called Nest Labs. Nest Labs was formed in 2010 by two former Apple engineers with the objective of bringing iPhone type of style and technology to the physical facility environment. Their first product was a smart thermostat that not only sensed temperature and communicated it over Wi-Fi to mobile phone apps but which also could "learn" the temperature preferences and habits of the humans in control of the thermostat. The Nest device would log for a few days of the human operator turning the temperature down at night and up during the day or up when the operator was away and back down when they returned to produce a history of the human operator's preferences. The Nest thermostat can then use this "learned" history of preferences and proceed to change the temperature in advance of the human operator's changes with the goal of relieving the human of the task of making future changes. The Nest thermostat did provide the option of deactivating this learning feature since there was usually more than one person who had an interest in what the thermostat was doing which made the simplistic learning feature not so helpful.

In the near future, there will be temperature sensing capabilities in almost any product or in any structure (such as a wall, window, ceiling, etc.) normally part of the interior of a home or office. Such sensors will provide a more localized and distributed set of data that can be used by smart air vent actuators to control the flow of heated or cooled air to only the location of the person in the facility instead of to the entire facility. Thus, we could have a personalized virtual thermostat system that provides conditioned air to a specific person as they move around a facility by monitoring where they are and by controlling the vents feeding each room.

Another example are the sensors that can be built into an automobile that can provide data on the movement of the vehicle and on the physical environment around it. This is at the heart of the "autonomous vehicle" or self-driving car. By integrating the data from radar or IR sensors all around a car, from speed sensors, from digital highway maps, from speed limit maps, and from routing plans, the control computer on a vehicle can provide hands-off control over the vehicle. Sensors are a key part of any autonomous vehicular system.

A summary of the evolution of sensor technologies is given in Exhibit 2.6.

EXHIBIT 2.6 SENSOR EVOLUTION

Sensor Technologies			
Mechanical	Electrical	Optical	Chemical
Acoustic frequency	Electric current	Infrared light energy	Catalytic reactions
Acoustic intensity	Electric voltage	Laser	Chemical-optical reactions
Density	Electrical capacitance	Light emitting diode	Heat flow
Fluid flow	Electrical inductance	Ultraviolet light energy	Ion reactions
Gravity	Electrical resistance	Visible light energy	Molecular reactions
Mechanical Force	Electromagnetic radiation		pH levels
Piezo-electric	Energy wavelengths		Phase changes
Pressure	Gravity		Redox reactions
Strain	Ionizing radiation		
Stress	Magnetic field		
Thermal expansion	Piezo-electric		
Vibration	Subatomic particles		

As wireless intelligent sensors continue to appear in everyday life, the Internet of Things will expand from 20 to 30 nodes per home, office, or vehicle to hundreds or thousands. Such increases in data generating devices increases the dimensionality of the data digestion that must be handled by automation by orders of magnitude.

Data

Data is just the digital representation of something that a sensor has measured, a computer has calculated, a memory device has stored, or a person has typed or said. Machines either collect, generate, process, and/or store data.

Data may or may not have value. Data that represents a patient's heart rate during surgery is highly valuable to the anesthesiologist and the patient. Data that represents the temperature of the interior of a garage at night in summer may not have value to anyone. It has been reported that at least 85% of data stored is dark (i.e. has unknown value), redundant, obsolete, or trivial (Veritas Technologies LLC, 2016) because it is stored in data centers and corporate or personal computers because few people or organizations have rational policies about how to value and archive any type of data. As more data is generated by the Internet of Things, this issue just gets larger.

However, many researchers and business analysts believe that if enough statistical analysis is done on enough data (e.g. Big Data Analytics), then value can be derived from massive amounts of seemingly useless sensor readings or stored internet transactions. The derived value could take the form of new knowledge about such things as consumer behavior, physiology, healthcare, physical phenomena, climatology, and engineering. In the physical and medical sciences, there is an increasing use of data analysis to accelerate the creation of new hypotheses or validation of existing hypotheses as an alternative to physical experimentation or human trials.

Early on, the value of data analysis was discovered by hedge funds and stock analysts as a way to model, estimate, or anticipate changes in stock prices. Although the Wall Street term used to describe a stock analyst with above average math skills was "Rocket Scientist", more sedate terms (data science and data scientist) are now used to describe the field and professional that applies statistical analysis to massive amounts of data.

There are also concerns that as more data is collected, individual privacy and national security become more vulnerable.

With respect to privacy, it is not clear what that term means anymore. Anyone who uses credit or communicates via electronic means (which is a majority of the world population) is exposing much of their daily activity to computers owned by business organizations and governments around the world. The term privacy is rapidly losing its meaning. The real challenge for most of us is how to manage our growing data footprint that we are generating each day.

With respect to national security, data is becoming one of the most important assets that can be weaponized and that must be protected. The major military organizations in the world have cyber warfare divisions whose purpose is to protect homeland data as well as attack the data of enemies and/or competitors.

What is the Digital Horizon?

The growth of the volume of data that is created, transmitted, and stored has grown exponentially in just the last 20 years. A summary of the volume growth is given in Exhibit 2.7.

EXHIBIT 2.7 GROWTH OF DATA MOVED AND STORED

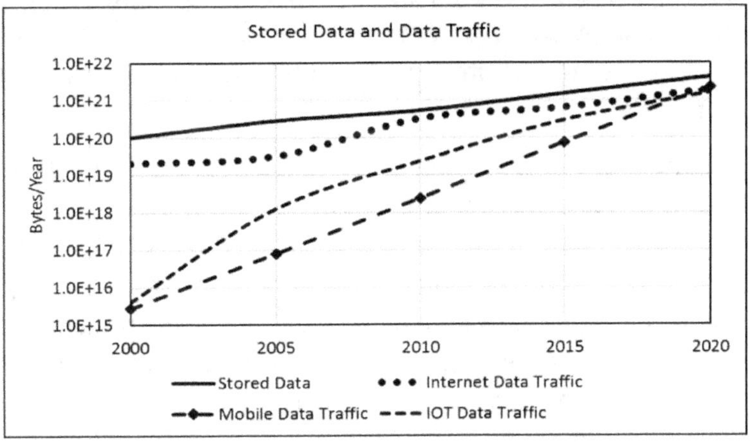

[Chart: Stored Data and Data Traffic, Bytes/Year from 1.0E+15 to 1.0E+22, years 2000–2020, showing Stored Data, Internet Data Traffic, Mobile Data Traffic, IOT Data Traffic]

Sources: Cisco, IEEE

Algorithms

An algorithm is a set of rules to be followed by someone or something performing a task. In the broadest sense, every computer program is an algorithm. The set of instructions being performed by the program could be the implementation of a complex set of mathematical analysis techniques for use with a set of data received or generated by a computational device. The instructions could also include evaluating, interpreting, and organizing data. Alan Turing invented the modern concept of the digital computer in the 1930's by using a logic algorithm in the disproof of a complex applied mathematical hypothesis ((Turing, 1937)

Ultimately, algorithms can be designed to make decisions and implement actions based on the data it has processed. Decisions made by algorithms can be for the purpose of controlling the actions of a machine, producing a report, or communicating with people or other machines.

A simple example of an algorithm are the instructions on the label on a bottle of shampoo that tells the user to: (1) Apply to wet hair, (2) Lather, (3) Rinse, (4) Repeat. A more complex algorithm would be a computer program running in a hedge fund data center that is monitoring the U.S. president's speeches, press conferences, and Tweeter messages for any mention of the name of a public company so that it starts trading stock in the referenced company ahead of human traders.

Innovation in the creation of algorithms has benefited from the invention of new programming languages that make it easier for a broader group of people in terms of skills and talents to convert ideas or policies into computer code. The very first computer programs were written in the machine language unique to each manufacturer of computer hardware, i.e. the very detailed instructions that a processor chip executes to move and change binary words. Then higher level languages (e.g. COBOL, FORTRAN, BASIC, ALGOL, C) were created that made it easier for humans to convert a set of specific mostly scientific or mathematical instructions into the machine language that the electronic computer hardware could execute. Later on, hundreds of additional languages (e.g. JAVA, HTML, C++, PHP, PYTHON, IOS/SWIFT) were created by corporations, research organizations, and industry consortiums to make it easier to implement instructions that dealt with data bases, graphic data, and other forms of data transmitted through the Internet.

Memory

Memory in the digital context is the storage of binary words. Alan Turing built pneumatic tube delay lines as his memory cells for the computer experiments he conducted prior to World War II. After WWII, several types of magnetic media were used with magnetic disk drives becoming the most economically viable mass storage technology. Since the age of semiconductors, solid state memory chips have become very dense and cost effective. A disk drive 30 years ago was the size of a refrigerator, used a kilowatt of power, and could only store a few hundred thousand bytes of data. In 2016, a solid-state thumb drive (the size of a thumb) can store gigabytes of data (millions of times more data in a device a thousandth the size). The data in Exhibit 2.8 indicates the rate of change in memory technology over the last several decades.

While there have been exponential improvements in memory cost and density, there is a long way to go before solid state electronic memory approaches that found in nature. Consider each nucleotide pair base pair on a single DNA strand in a human cell as a bit (binary digit) of data. This translates into a storage capacity for one human cell of roughly 3,000 Mbytes and a storage density of 6.0×10^{14} Mbytes/in³. The comparison: one human cell has about 10^{12} (a trillion) greater density than the best memory device that man has been able to build.

Making future automation more effective depends on continuing the reduction in the cost of and increasing the density of solid state memory devices. It has been hypothesized that biologic based or quantum based technologies could be successors to the silicon based memory devices.

EXHIBIT 2.8 DIGITAL MEMORY TRENDS

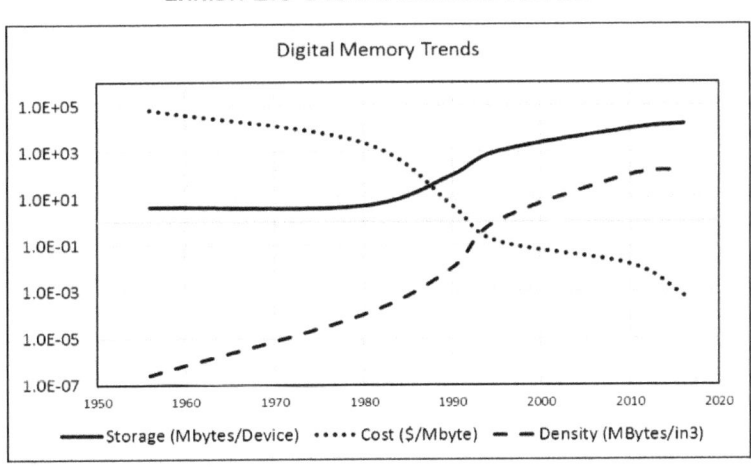

Sources: IBM, Seagate, Western Digital

Power

Machines require power to operate. If a machine is not tethered to an electrical outlet for power, then it must carry its own power source along with it. Fossil fuel powered machines (e.g. tractors, cars, trucks, ships, planes) carry the fuel along with the machine. Battery powered machines have limitations in weight, range, and operational duration because of the low energy to weight ratio of conventional battery technologies.

Battery technology has not made as many orders of magnitude improvement over the years as memory technology (Exhibit 2.9). Where data storage density has improved by a factor of 10^8 over the last 4 decades, battery energy storage density has only increased by a factor of 2 (comparing lithium ion batteries to lead acid batteries). Where data storage cost has decreased by a factor of 10^{-8} over the last 4 decades, energy storage cost in a battery has decreased by only a factor of 10^{-2}.

The cost and size of power storage are two of the reasons that robotic applications for consumer applications have not become practical yet. Before robotics can make a noticeable inroad into the general economy, the size and weight of the mechanical portions of the machine must be reduced further to match the size and capacity of stored energy devices.

EXHIBIT 2.9 BATTERY TRENDS

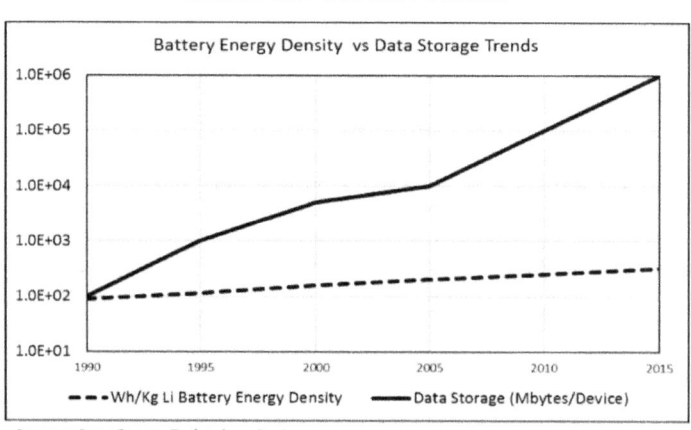

Sources: SqurrEnergy, Technology Review

Systems and Synergy

Typically, all this digital technology has value only when it is part of a system. A system is a collection of some combination of the fundamental technical elements described above and the interfaces with other systems and with humans. Creating an architecture for a system requires just as much creativity and is as critical to innovation as any of the individual elements.

An automobile is a system. The dashboard display in the automobile is a system. The audio features in the automobile is a system. In the typical automobile made in 2016, there are hierarchies of more the 50 systems and subsystems with more than 100 semiconductors acting as controllers or sensors.

On the other end of the scale, the Google "cloud" is a global system of data centers which house a system of servers (e.g. a single computer board with many terabytes of memory). Each data center houses more than 20 million servers and there are more than 100 Google data centers around the world. There are more than 20 companies that have more than 10 data centers each.

One of the basic structures of electronic systems is the printed circuit board (PCB). Invented over 60 years ago at the dawn of the electronics age, a PCB is typically a plastic (although other substrate materials such as glass and ceramics are finding applications) board on which flat wires of conductor (usually copper) are deposited to provide connections to electronic components (such as microprocessors, memory chips, light emitting diodes, etc.) have been mounted (see Exhibit 2.10).

The conventional PCB is flat, rigid, and rectangular in shape. It may be composed of multiple laminated layers of conductors with the electronic components attached on the top and/or bottom outside layers. One of the components on a PCB is usually a microprocessor which is itself a printed circuit on silicon. The ubiquitous "mother board" of personal computers is a PCB which has the primary microprocessor on board.

As shown in Exhibit 2.10, printed circuits have evolved to flexible plastic with the thickness of cellophane. Cell phones, tablets, laptops, and wireless remote controllers are chock full of flexible PCBs (often called flex circuits). Flex circuit dimensions are beginning to approach the dimensions of the semiconductor technology from decades ago. Flex circuits can be made small enough to fit on the end of intravenous catheters and carry active components such as ultrasonic scanners and into pill sized capsules for medical diagnostics sensing.

EXHIBIT 2.10 PRINTED CIRCUIT TECHNOLOGY EVOLUTION

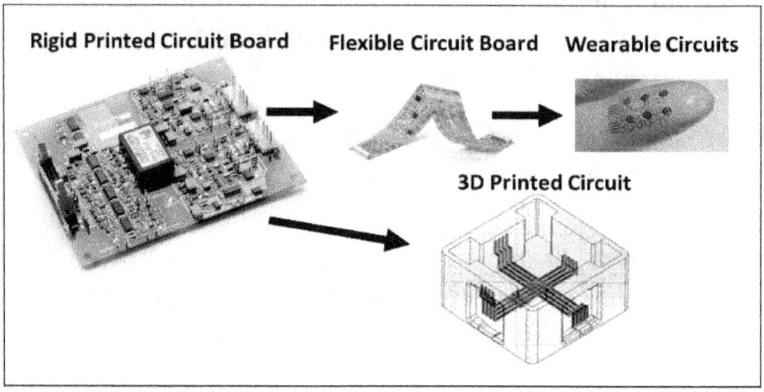

Rigid Printed Circuit Board → Flexible Circuit Board → Wearable Circuits

3D Printed Circuit

Sources: Jabil, Flextronics, Samsung

Other new applications for printed circuits that are causing innovation in architecture include wearable electronics and 3D printed devices. In wearable electronics, flex circuits are integrated into wrist bands, eyeglasses, adhesive patches, and clothing for applications such as medical monitoring, entertainment, physical fitness, and mobile communications. In 3D printed devices, the conventional three-dimensional Cartesian PCB structure made of laminated plastic sheets is replaced by printing conducting metal paths within and throughout a 3D printed part. A coffee cup could be created with a 3D printer which has embedded in the walls and the floor of the cup enough circuitry and components to make the cup an intelligent networked device.

Systems technology is evolving to include not only the conventional electronic components but also the clothes we wear and the everyday items we use. As the number of everyday items that become smart (have digital technology on board) grows, the importance of networking continues to increase. In terms of network nodes (or communicating points in the network), the IoT is already larger than the Internet.

And as the networked world expands in scope and geography, the concept of decentralized architectures for systems becomes more important. The conventional Internet and the emerging IoT are inherently decentralized. In the digital world, decentralized systems and architectures are already making inroads in entertainment (streaming of audio and video content, self-publishing video and literary websites, etc.), in home automation (Wi-Fi controlled thermostats, sprinkler systems, door locks, electrical outlets, appliances, etc.), in police and safety (traffic violation cameras, highway cameras, smart stoplights, etc.), in industrial systems (supply chain automation, distribution center and warehouse management, retail stores, etc.), in power generation systems (roof top solar, wind turbines, battery storage systems), and in military systems (drone weapons, smart ammunition, sea-based and land-based motion systems, etc.). The highly anticipated world of autonomous vehicles will consist of a national highway network of decentralized systems on 4 wheels.

The extent and amount of economic benefit that will be produced by innovation in the technology of systems will be determined in large part by the amount of synergy that emerges. A smart pill that can measure a patient's medical status or precisely deliver medicine will have limited benefit if the healthcare system that wants to deploy it has not adapted its processes and payment systems. If it becomes commercially practical to pack a chemistry lab, pharmaceutical dispensary, or diagnostic system into a smart pill, a significant portion of the conventional health care system will be disrupted. And if there are medical AI systems developed to process the data coming from smart pills, then other portions of the healthcare system will be disrupted.

It is becoming clearer that the big legacy centralized system structures that have been such an important part of the successful productivity and economic growth of the last century will become liabilities as new decentralized, highly intelligent, and highly networked digital systems emerge. Disruption in key industries in our economies is inevitable and necessary as major innovations emerge that improve productivity, create value, and redefine how duties are shared between people and machines.

SECTION II

ECONOMICS: If No ROI, Then No AI

Chapter 3

THE PRODUCTIVITY CYCLE

Humanity's Drive to Be Better

Based on a variety of measures, it appears that evolution has sharpened one human trait: the innate drive to improve. Whether it be in the availability and variety of food, the quality and range of health care, the size and utility of housing and transportation, or in the variety and availability of entertainment, mankind has found ways to improve, i.e. to do better than previous generations. In just the last 100 years (Exhibit 3.1), the average lifespan has increased, the average level of nourishment has increased, the average level of wealth has increased, the average years of education has increased, etc.

EXHIBIT 3.1 EXAMPLES OF HUMANITY'S IMPROVEMENTS

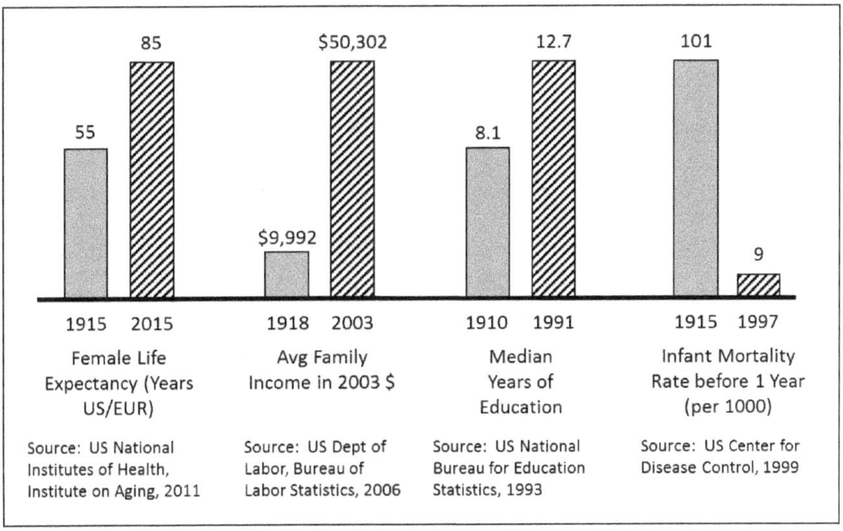

1915	2015	1918	2003	1910	1991	1915	1997
55	85	$9,992	$50,302	8.1	12.7	101	9
Female Life Expectancy (Years US/EUR)		Avg Family Income in 2003 $		Median Years of Education		Infant Mortality Rate before 1 Year (per 1000)	
Source: US National Institutes of Health, Institute on Aging, 2011		Source: US Dept of Labor, Bureau of Labor Statistics, 2006		Source: US National Bureau for Education Statistics, 1993		Source: US Center for Disease Control, 1999	

If the main hypothesis of the theory of evolution is survival of the fittest, then the human species has learned how to not only be fit enough to survive amongst competing lifeforms, but to also prosper and improve its wellbeing beyond anything that random genetic variations might produce. The uniquely human traits of intelligence, innovation, and initiative have resulted in humanity doing more than just out-hunting or out-grazing the other competing species in the wild.

One model to indicate how innovation and initiative might factor into our social and physical changes is Maslow's hierarchy of needs (Maslow, 1954) (Exhibit 3.2). Maslow's hierarchy is heuristically attractive for defining what might be motivating a person, how their initiative is being applied, or what type of reward they are seeking. For one person, it might be enough (or maybe the only path available) to advance from the physiological level of satisfaction to the safety level. For other people, seeking a higher level on the model might be the only goal so that safety or belonging might not be sought or achieved. The point of mentioning this model here is not to get into a psychological discussion about the efficacy of the model, but to use it to highlight that we as a species drives to be better off in the future than we were in the past.

Exhibit 3.2: Maslow's Hierarchy of Needs

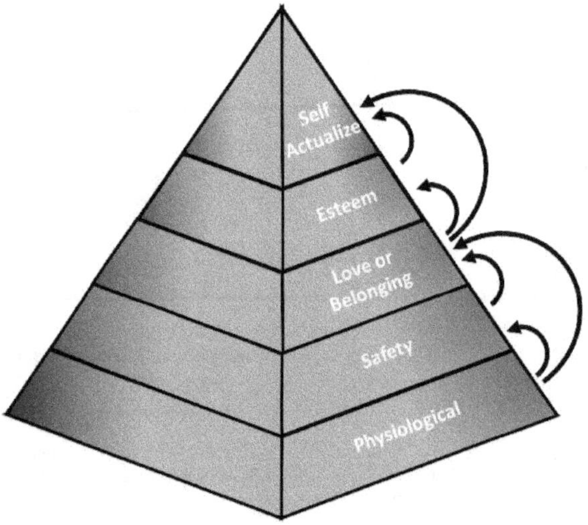

The definition chosen for being better here is getting better economically. Being better off economically generally means people have more wealth, can afford more leisure time, better healthcare, and more education. How much better an economy gets can easily be measured by one key parameter: productivity. Being more productive means that a person, machine, animal, or organization can produce more output with the same or less input. More productive labor means that more goods or services can be produced with fewer hours of labor. More productive machinery means that more goods or services can be produced with fewer, smaller, or cheaper machines. More productive capital means that a greater value of goods and services can be produced for less cost.

While economists clearly state that productivity improvement is necessary for an economy to grow and for its citizens to prosper [e.g. (Federal Bank of Dallas, 2016), (Shakelton, 2013)], what the causes and effects might be that lead to productivity improvement in an economy has been under study for decades by academics and government agencies. Rather than report on an analysis of the academic work that has been done on this topic, two simple models are presented in the following to illustrate how productivity improvement tends to happen.

The first model is the Productivity Improvement Model (Exhibit 3.3). It is a cyclical model composed of five key areas each of which is described below. The second model is the Innovation Cycle and its interaction with the Productivity Improvement Model is discussed in Chapter 4.0.

EXHIBIT 3.3 THE PRODUCTIVITY IMPROVEMENT CYCLE

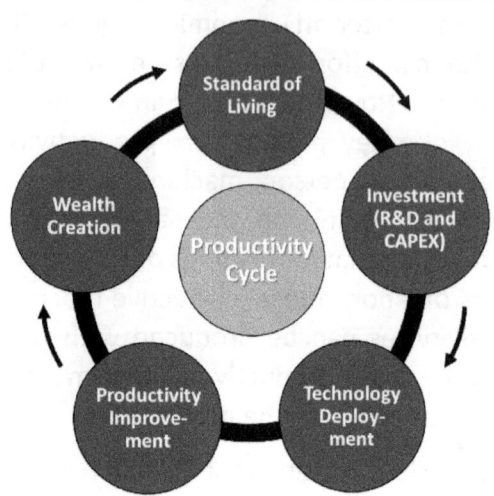

Standard of Living

Standard of living refers to the typical economic well-being of a group of people. Economic well-being includes amount and type of material possessions, level of comfort, availability of food and entertainment, availability of health care, and quality and availability of housing. Intangible factors such as level of happiness or job satisfaction are not included in this definition of standard of living. Metrics for standard of living are usually created to make it possible to answer questions about changes in time or geography. How does the standard of living today compare to 50 years ago? How does the standard of living of the US compare to Norway? Examples of simple metrics for standard of living include GDP per capita or disposable income per capita.

But there are other metrics that provide insight into the standard of living for a state or nation, such as (1) savings as percent of household income, (2) taxes as percent of gross income, (3) food cost as percent of household income, or (4) healthcare cost as percent of household income. For the purpose of this productivity model, these metrics can be useful.

Standard of living has been included as an element of the Productivity Model because it captures the concept of improvement as alluded to in Maslow's hierarchy of needs. Given any level of standard of living, there is usually an upwelling of new unmet needs that may not have be considered possible or even conceivable before. This upwelling occurs either because new levels of wealth create new demands for previously unaffordable things or because someone has come up with a new value-creating innovation that previously did not exist. In any case, whether a new need is obvious and well-articulated (e.g. a cure for breast cancer) or an consumer economic surprise (e.g. smart phones), it is a need that can spur economic improvement.

The next question is how are the needs for improvement identified or communicated to the organizations or people that can do something about it? In most of the economically developed nations, this communication takes place in some form of efficient market mechanism between corporations, consumers, governments, and media. In totalitarian governments or underdeveloped nations, market mechanisms seldom exist or are inefficient because of bureaucratic or cultural barriers. In economies where there are effective market mechanisms, there is enough exchange of needs and ideas and information for investors (private and/or public) to act by investing in the factors that make economic productivity improve. In economies where there are artificial or cultural barriers, investment is not encouraged or is unattractive resulting in minimal or negative economic progress.

Investment

Investment is a term that can be used to describe a variety of activities where money, time, or labor is applied to achieve a long-term goal. Investments can be made through the financial markets (stocks, bonds, funds, certificates of deposit, savings accounts), through real estate markets (property ownership), or through direct involvement in new value creation. In this treatment, the focus is on three categories (Exhibit 3.4) of the new value creation investment each with a different risk profile and each treated differently from an accounting and tax perspective.

The largest category is capital expense (CAPEX) which describes money invested by private and public business enterprises in the creation and implementation of new technology, facilities and equipment. CAPEX is used by companies to expand their asset base to support or stimulate revenue growth, increased market share, and hopefully improved profit margins. CAPEX examples include new automation on an automobile production line, new assembly robots on a mobile phone production line, or new information systems for a hospital. The time frame for an initial positive return (some combination of cash flow and capital valuation) from CAPEX investment is usually planned to be five years or less. The typical success rate (all project objectives are met or exceeded) for corporate CAPEX projects tends to fall in a range of 50% to 75%.

EXHIBIT 3.4 US VALUE CREATION INVESTMENT TYPES

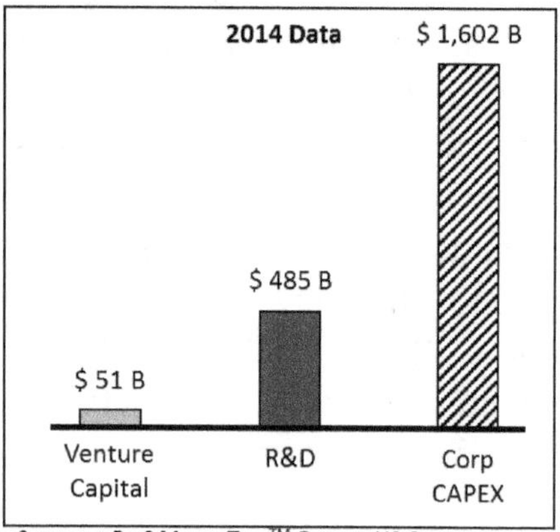

	2014 Data	
Venture Capital $51 B	R&D $485 B	Corp CAPEX $1,602 B

Sources: PwC MoneyTree™ Report, US Census Bureau, R&D Magazine

A second category of new value creation investment is research and development (R&D). R&D includes money spent on new discoveries, inventions, and product development. The sources of R&D spending in the US are identified in Exhibit 3.5. R&D spending tends to be a higher risk investment because of the uncertainty associated with the outcomes from the research and development activities. Not every new product development project is met with success in the marketplace. Not every invention proves that it has value to future users. Not every search for a scientific discovery results in new useful knowledge. The time frame for an initial positive return (again some combination of cash flow and capital valuation) from R&D investment is usually hoped to be in 10 years or less although some life science R&D projects can have a time horizon of a decade or more. The typical success rate for R&D projects tends to fall in a range of 10% to 20%.

EXHIBIT 3.5 SOURCES OF R&D SPENDING

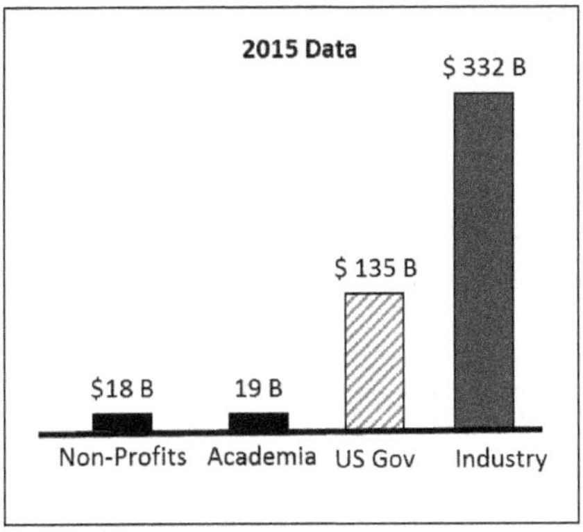

2015 Data

- Non-Profits: $18 B
- Academia: 19 B
- US Gov: $135 B
- Industry: $332 B

Sources: R&D Magazine

A third category is investment in the creation of new business ventures (Venture Capital). This is the highest risk category of the three considered here because it involves the creation and implementation of new technology in addition to the creation of a new business enterprise to take the new technology to market. The time frame for an initial positive return (typically only capital valuation that is monetized from an IPO or an acquisition by a larger company) from a venture capital investment is usually hoped to be in 3 to 7 years. The typical success rate for venture capital investments tends to fall in a 2% to 5% range.

The sources of investment money include corporations, governments, non-profits, financial organizations, and individuals. Money for investment comes from a portion of profits earned by corporations, from savings of individuals, from wealth accumulated in non-profits, and from taxes collected by governments. Much of the money from individual's savings, non-profits, and some corporations flows into investments through the financial markets. Corporations flow most of their CAPEX investments through their own subsidiaries and their contractors. Governments flow their investments through government owned agencies to universities or independent research organizations.

Money for investment comes from domestic sources (internal to a nation's boundaries) or from foreign sources. In fact, metrics of the economic "health" of a nation include whether its citizens are putting a large enough percentage of their disposable income into savings and whether there is money from other countries being invested inside its borders in new value creation projects. If a country is not generating enough savings or attracting enough foreign money for investment, there is little chance that the nation's economy can improve or that the economic wellbeing of its citizens will improve.

Technology Deployment

Technology deployment is a critical phase in the productivity improvement cycle. Investment money facilitates the creation of new technology and the launch of its introduction into the market. Deployment is important because it represents the movement of a technology out of a laboratory or engineering office into the hands of paying customers. It starts the generation of new revenue and begins a new flow of cash to the original investors which usually leads to the creation of new capital value.

Technology deployment is nontrivial and in many ways can be more challenging than the discovery, development, and creation of the technology. It also takes time as well as money. And the amount of time and money depends on the readiness of the marketplace to accept it and the cleverness of the people rolling it out.

One direct example of the cost and time to deploy technology is semiconductors. When Intel develops a new microprocessor that runs faster, is smaller, and costs less than anything that preceded it, it has to invest several billion $US in a new highly automated manufacturing line that takes up to 2 years to build and bring online.

However, as soon as the new semiconductor fab begins making parts, if Intel was correct in its assessment that the new cheaper faster microprocessors are needed, then revenue begins to be generated with each new microprocessor sold.

A less direct but equally relevant example is online search engines. When Google was founded 20 years ago, there were dozens of Internet search engine services available. Few of the competing search engines generated any revenue. Also at that time, search engines were able to find less than 10% of the pages on a very small Internet that had less than a few thousand web pages. However, Google not only had invented an efficient new algorithm for prioritizing search results (the Page Rank method), they wanted to find and link to every web page on the Internet. This meant that the new company had to have the ability to store and retrieve key information about every page on the observable Internet. Google also believed that high speed response was essential for the quality of their service. This strategy led Google to realize that they would need to have high speed data centers to store all of the addresses of all the page links with key words so that their search customers could get results in just a few milliseconds. This was a level of service unseen before among search engines. Since Google's founding in 1996, the Internet has grown from a few thousand pages to trillions of pages (growth of 1,000,000 X), and Google's has dozens of data centers around the world housing millions of server computers and representing many billions $US in investment. All for a service that Google gives away free. Of course, the large investment required to create the networking, storage, and processing power would not have been made if Google had not also invented a business model to use the service as a vehicle for selling directed advertising. The investment in Google not only disrupted the search engine industry but also the advertising industry.

The point here is that it is no matter how interesting or compelling a new technology development might appear, if there is not enough market demand and deployment funding to get the technology to market, then the new technology will sit on a shelf until time passes it by or until the market recognizes its value and need.

Productivity Improvement

Once the investment is made to finish the development of a new technology and then to get it deployed into the market, the hope is that it should start generating a financial return that rewards the investors that risked their money in the first place. While market hype or industry buzz might be enough to generate an initial surge in interest on the part of customers, if there is not a sustainable financial or economic benefit, then there will likely not be a successful or enduring use of the technology. Productivity improvement is essential for the establishment of such a sustainable financial return for the investors.

While historical examples are highlighted in several places in this book, modern examples abound. Often the first thing that one might think of when the words "productivity improvement driven return on investment" are mentioned is a manufacturing industry. However, productivity improvement in service and information industries are just as or even more important than manufacturing industries in modern economies.

Consider the advertising industry for example. Google's financial success can be attributed to the greater productivity of advertising money spent on AdSense compared to more traditional advertising channels such as print media and broadcast television and radio. With online advertising, the probability of reaching a target audience or demographic segment is significantly higher. In 2014 Proctor and Gamble, the largest single spender of advertising money in the US, reduced their total advertising budget by 4.2% but announced they were getting a greater return on their advertising investment by moving more of their advertising placements to digital media such as search, social networking, video, and mobile. Their data confirmed that consumers are spending more time on such media and less time on newspapers, radio, and television (Johnson, 2015).

While productivity improvement in one industry might mean the elimination of jobs, it also usually means the creation of new jobs in other industries that may or may not be directly related. In Exhibit

3.6 a comparison of advertising revenue and job decline in the print industry is compared to the increase of advertising revenue and job growth in the online search industry.

EXHIBIT 3.6 ADVERTISING INDUSTRY TRANSFORMATION

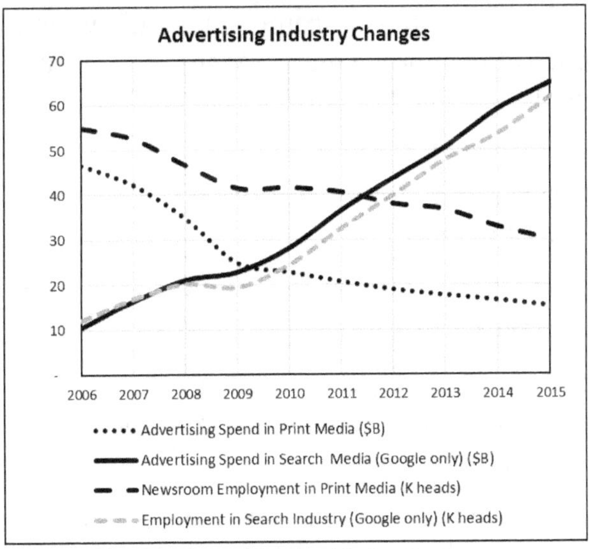

Advertising Industry Changes

····· Advertising Spend in Print Media ($B)
—— Advertising Spend in Search Media (Google only) ($B)
– – Newsroom Employment in Print Media (K heads)
– – – Employment in Search Industry (Google only) (K heads)

Sources: Advertising Age and Google public documents

While the productivity of a dollar of advertising spending in the online search industry was increasing, the amount of advertising spending and the number of employees in the print media was decreasing. The data also shows that the number of new jobs created in the online search industry was approximately twice the number of jobs lost in the print media. While it should be noted that many of the types of jobs created were likely to have required different skill sets and types of training, there were some that had similar requirements. The jobs lost in the print media that did not translate to the online industry included printing press operators, newspaper delivery people, and typists. The jobs created in the online search industry included data center operators, software engineers, and data analytics experts. Some job types (such as inside sales agents, account managers, and accountants) did translate from the print media to the online search industry.

Wealth Creation

New wealth is created after investment has been wisely made, technology successfully deployment, and productivity improved in the creation and delivery of one unit of product (or unit of service) or in the performance of one company, industry, and/or country. Since productivity improvement means that more output is generated with less input, then one of the following phenomena follows:

1. If prices stay the same, then more profit margin is generated (because cost of units sold is less)
2. If prices decrease, then more units are sold (more affordable units increases demand)
3. If more units are sold, then market share is taken from the competition (assuming competition has not achieved the same productivity improvement)
4. If the rate of unit sales increase is greater than price decrease, more revenue with an improved profit margin is generated (more units sold at better profit margins)
5. If prices decrease, then customers either buy more units or have new funds to spend on other products and services

In each of these five outcomes, good things happen in the economy and new wealth is created.

In the case of phenomenon #1 above, more profit margin should translate into greater net profit for the company which means that more value is created for the shareholders, more cash is available to the company to invest in other growth opportunities, and more cash can be returned to the shareholders, investors, and employees.

In the case of phenomena #2 and #3, more unit sales translates into more market share which should translate into more pricing control or more economies of scale which increases the opportunity to increase gross profit.

In the case of phenomenon #4, increased revenue at an improved profit margin should translate into more net profit.

In the case of phenomenon 5, there is economic benefit to other parts of the economy beyond the benefit to the company that achieved the productivity improvement because new money has been made available to the customer that should increase buying power or increase the money that the customer save.

Wealth is a term which connotes tangible assets but in reality often relates to intangibles. Bill Gates is always at or near the top of Forbes richest person list with an estimated worth of $100 billion mostly based on the number of shares of Microsoft Corporation that he owns times the price of the shares at the snapshot in time when the list is made. However, if Mr. Gates decided to sell all of his shares in Microsoft at once, the price of the shares would likely drop precipitously because the demand for his shares would be considerably less than the supply that he is selling. Every time that the Microsoft share price plunges on the public stock markets, there are headlines touting how many billions of dollars of wealth Mr. Gates lost in the span of one trading day.

Wealth refers to the value of something that a person or organization owns. For investors, it is the ownership of shares of stock. For creditors, it is the ownership of debt. For employees, it is the value of the cash income from a job or the value of their savings. For individuals, it is the value of the cash income from a job, the value of their savings accounts, the equity they have in their home, the market value of jewelry, art, or financial accounts. For corporations, it is the ownership of assets and retained earnings. For wealth to increase, the number of things that a person or organization owns has to increase or the value of those things increase or both. Therefore, wealth creation is the creation of new value and/or the creation of new things that have value. The fruits of productivity improvement contribute to both.

As wealth increases in an economy, there is more wealth to taxed by governments, more wealth to be saved and invested in new economic opportunities, more wealth to be spent on consumables or intangibles or lifestyle. In an aging population, more wealth to be spent on healthcare and health maintenance. In a developing economy, more wealth to be spent on infrastructure and community services. More wealth in the economy should translate into an improved standard of living.

Law of Diminishing Returns

The Productivity Improvement Cycle has one major limitation that must be overcome: the law of diminishing returns. The law of diminishing returns is a well-established principle in economics theory that states that if one input in the production of a commodity is increased while all other inputs are held constant, a point will be reached at which the benefits from additional inputs become smaller and eventually diminish to nothing (Encyclopaedia Britannica, 2016).

A simple example of diminishing returns would be a farmer who owns 100 acres of farmland on which corn is planted (see Exhibit 3.7 and references (University of Illinois, 2016) and (Iowa State University, 2016)). When the corn has grown to the point it is ready to be harvested, the farmer has 5 days in which to harvest the corn before the crop begins to spoil. One acre can produce about 175 bushels of corn. One field worker can harvest and shuck about 50 bushels a day. Thus it would take 350 field hands to harvest all the corn in one day or 70 field hands to harvest in 5 days. The total cost of harvested corn is $2.00 per bushel.

EXHIBIT 3.7 EXAMPLE OF DIMINISHING RETURNS

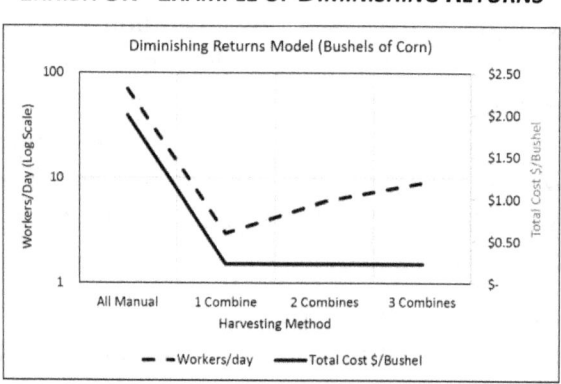

With automation (using a modern 500 horsepower combine harvester), the number of field workers is reduced to 3 and the total cost per harvested bushel is $0.22. Adding a second combine to the process reduces the harvest time by 3 hours but does not reduce the cost per bushel. Adding a third combine again reduces harvest time actually increases the need for personnel but does not reduce the cost per bushel any further. The point of diminishing returns for this farmer is one combine.

Of course, a year or two after the farmer purchased his 500 horsepower combine, an enterprising combine salesman might tell the farmer that a 600 horsepower combine has come on the market that was 50% wider and would therefore reduce the cost per bushel even further. Even with a favorable financing plan plus a generous trade-in, this would not be considered an innovation. The point of diminishing return is still one machine.

A more generic view of how the law of diminishing returns affects productivity improvement is illustrated in Exhibit 3.8. This curve shows that productivity improvement decreases to zero (productivity curve flattens out) after some level of capital investment. The message is that no additional financial benefit will occur after a certain level of capital investment is made.

EXHIBIT 3.8 PRODUCTIVITY IMPROVEMENT AND DIMINISHING RETURNS

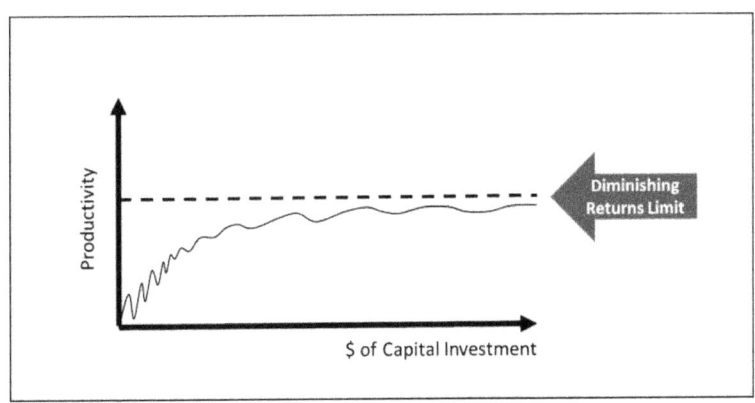

Of course the operative assumption with this economic hypothesis is that everything else remains the same: that there has been no significant progress in the people, technology, or processes that are being used. If this assumption were reality, then once the point of diminishing returns is reached, an economy would not grow and no new wealth would be created. If the population continued to grow, then output per person would flatten or decline and productivity change would become negative.

This type of economic performance would describe the reality of the US and most developed economies over the years 2010 through 2016 (Exhibit 3.9). During this period, the productivity of labor in the US grew only about 0.3% per year (historic lows) until 2016 when it actually declined (Soergel, 2016). Nonresidential fixed investment, considered an indicator for how much money domestic businesses were putting back into their own operations fell during 2016 as did investments into new equipment and spending on new structures. At the same time, new jobs were being created and unemployment was falling. The productivity ratio was turned upside since the economy was doing less with more. This was happening despite years of economic stimulus from budget deficits at the federal government level and money creation by the Federal Reserve.

EXHIBIT 3.9 US LABOR PRODUCTIVITY HISTORY

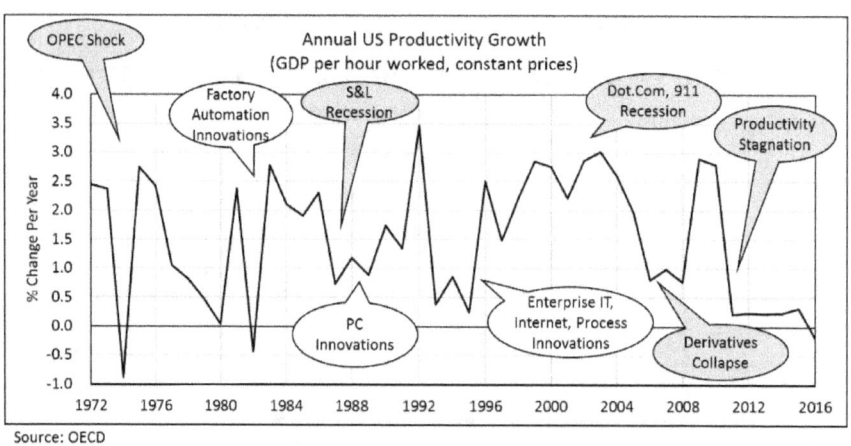

Source: OECD

As history has shown, periods of productivity decline can be overcome with the interjection of new technology, new training and education, and more efficient business processes. Innovation in any one or all of these areas can cause a sudden jolt in productivity improvement which puts the productivity curve on a new growth path (Exhibit 3.10).

EXHIBIT 3.10 INNOVATION OVERCOMES DIMINISHING RETURNS

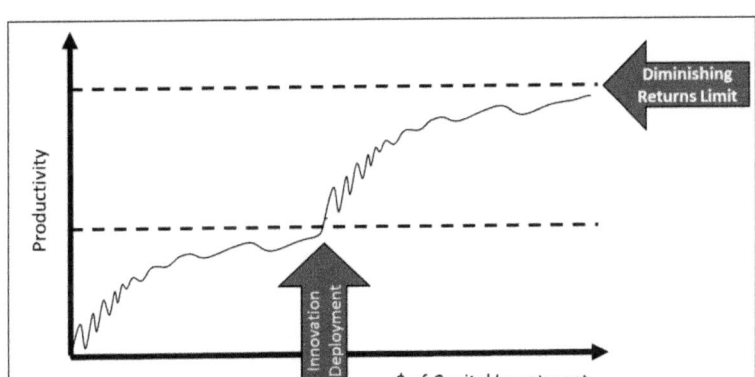

The drop off in productivity over the last five years happened despite major innovations in areas such as smart phone technology, exponential increases in Internet pages, and new forms of streaming entertainment. It appears that several of the recent technology innovations have reached saturation points with value creation diminished to near zero. There were more cell phones than people, more laptops than office workers, and a growing amount of digital storage media (disk drives, memory chips, data centers) purchased and dedicated to storing likely useless data files and videos created by smart phones, social network users, cloud computing, and GoPro camera. Clearly there is a need for a new wave of innovation to prevent economies from going into further economic stagnation or decline.

Chapter 4

THE INNOVATION CYCLE

Innovation can be described as a cyclical model separate (Exhibit 4.1) from the productivity cycle. The productivity cycle model is all about getting the necessary investment to deploy the available technology into the economy so that new wealth can be created. But as discussed above, without innovation, productivity improvement inevitably reaches a diminishing point and productivity would eventually flatten or more likely decline.

EXHIBIT 4.1 THE INNOVATION CYCLE

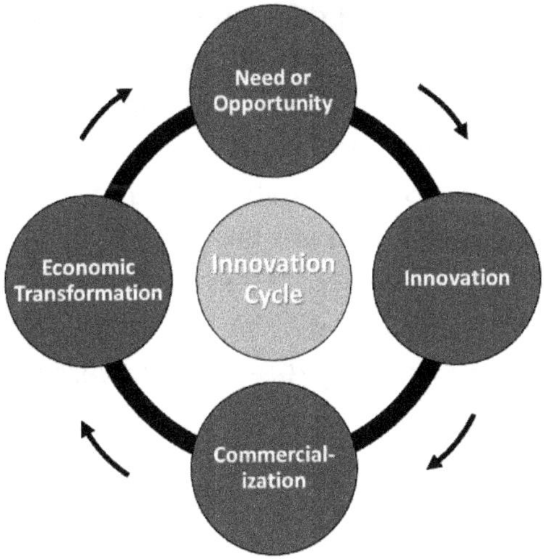

Need and Opportunity

The top of the Innovation Cycle is the existence of a need or opportunity for an innovation. Needs are often created by economic or social problems: the need for a new treatment of a disease, a new insecticide for a crop pest, a new paint to prevent rust, or a new energy source to reduce carbon emissions. Opportunities are created when an innovation occurs that creates a need that did not exist beforehand. Examples include social networking software (Facebook) to connect people instantly, a miniature digital jukebox that carries 1000 songs in an electronic package (iPod) the size of a matchbox, a smart phone (iPhone) that combines a camera, phone, music catalogue, and video player, or, a new video camera (GoPro) that you can wear on your hat while biking. In either case, whether it is need "pull" or opportunity "push", there has to be a stimulus behind the inspiration that leads to innovation.

Innovation

Innovation is the act of creating new technology, processes, products, or knowledge to satisfy a clear market need or to stimulate, nurture, and exploit a market opportunity. Innovation can be chaotic, organized, or accidental. It can be pursued by professionally managed research organizations, by entrepreneurial groups of inventors, or by focused and dedicated individuals. Innovation can be a clever way of doing the same thing but in a different but more efficient way. Innovation can be a major breakthrough resulting from a scientific discovery or a revolutionary new system design. It can also be an obvious combination of things that many people hypothesized but no one had yet implemented. Innovation can be the reinvention of something that had been invented before but for which the market was not ready. Innovation can be the offspring of necessity (to reinvent the old expression: necessity is the mother of invention).

Commercialization

Commercialization is the act of taking an innovation to market. An invention sitting on a lab bench or stored away in disk drive creates no value. In fact, the best definition of an innovation is an invention that has been successfully taken to market. If an invention is not creating value, is not generating revenue in the economy, or is not creating a benefit, it just an invention and not an innovation. There are millions of patents that have been issued since 1790 and only a small percentage of them have led to value creating, market successful innovations.

Commercialization usually involves taking a new technology, product, or process and building a business around it. The technology for a smart phone had been in existence for over 10 years, but it was not until Apple created an entire ecosystem for it (iTunes, App Store, Maps, Calendar, Contact List, etc.) that the real value of the innovation was unleashed in the cell phone market. The invention of electric power for consumers did not become a commercially successful innovation until the economics of alternating current for the local transmission of power was proven to be superior to that of direct current. The invention of personal computer technology did not become a significant innovation until spreadsheet application packages made it a necessary tool in offices and schools worldwide.

Commercialization does not happen without investment. In the US alone in there was over $58B invested by venture firms (PwC, 2016). The investment by public and private corporations in the commercialization of their internal research and development and new product development activities is more than 10 times that amount.

The investment required for commercialization includes more than investment in just research and development. For a new product or service to achieve market success requires understanding what the market wants and needs, how to give the market a platform for buying and receiving the new product or service, and how to create a supply chain that supports the creation and delivery of the new product or service. Successfully commercializing an invention, new idea, or improvement so that it becomes an innovation requires investment that goes beyond just R&D.

One might say that this definition of commercialization sounds like all the usual management things that a typical business should be doing. And that would be a correct statement. The difference is that an innovative product or service is disruptive to the market status quo. Innovation is disruptive by causing the market to think differently about how existing products are being used or how money should be spent. The concept of a new product or service might be disruptive in theory, but turning the concept into market reality usually requires the business skills of speed, flexibility, scalability, and messaging for which not every company has competency.

Economic Transformation

Economic transformation is the result of a successful commercialization of a new technology, product, or process. A transformation can be a disruption in an established industry or in the creation of new industries. Indicators of a transformation include sudden significant improvements in productivity, rapid creation of new value, formation of new ventures, or merger and acquisition consolidations in established industries.

The music industry, for example, has seen a series of economic transformations all driven by digital technology. The digitization of recorded music began when the consumer electronics industry launched compact disc (CD) players in the 1990's. The published music industry saw a surge in sales from new CD's with an average price of $15 being sold to a large volume of mobile CD playing platforms (boom boxes, automobile CD players, wearable CD players, etc.).

However, the DOT.COM bubble in the late 1990's stimulated the creation of many new Internet start-ups, one of which was Napster. Napster was a peer-to-peer digital music file sharing service which meant that the digital music files formerly sold on CD's suddenly could be copied and shared for free across the entire Internet. The value of a digitally recorded album suddenly went from $15 to $0. All of the physical distribution and media barriers that helped protect the music industry were made useless in just a few months. The only barriers that remained were legal barriers in the form of intellectual property ownership and distribution rights contracts. Napster was sent into bankruptcy after a series of legal battles but not before the music industry had been permanently changed.

Apple brought some stability to the music industry with its introduction of the iPod (and later the iPhone) and a negotiated relationship with music publishers of royalty payments of a few cents per digital file sold on its iTunes platform. As shown in Exhibit 4.2, the sales of single and album units on CDs peaked shortly after the online company Napster was founded with the decline continuing after Napster shut-down and the launch of iTunes and iPod. The units downloaded has grown rapidly since the launch of the Apple products and is now almost twice the peak of the CD units. Meanwhile, CD unit sales are diminishing to likely the same fate as cassettes. The sale of mobile digital devices that can store and play thousands of music files has grown to exceed 7 billion units and is greater than the world population.

EXHIBIT 4.2 MUSIC INDUSTRY SALES IN UNITS

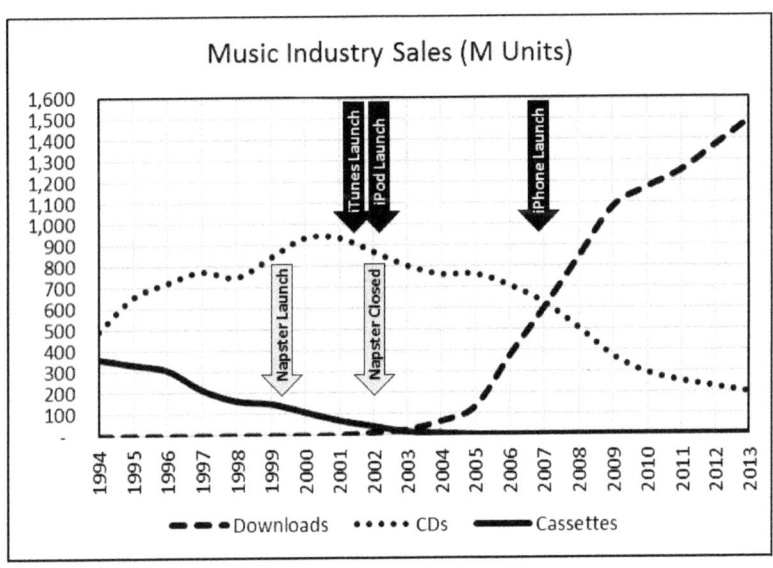

Source: Recording Industry Association of America

The economic transformation of the music industry affected not only the publishers and distributors of music but also the economic factors for the talent (authors and performers) in the industry. During the era of the cassettes and CDs, performers would engage in multi-city tours to encourage their fans to buy those tangible products. In the era of digital streams where there is no physical product and revenue comes only from a few cents per play or download, performers engage in multi-city tours for large gate receipts. In CD era, performers gave concerts to stimulate CD sales. In the streaming era, performers nearly give away their streamed music to generate ticket sales at their concerts.

The economic transformation of the mobile device industry created new needs and opportunities for digital music which led to the transformation of the music industry. The economic transformation of the music industry to go to digital downloads and streams created new needs and opportunities for live performances.

In summary, the economic benefits that accrue from each economic transformation leads to new ideas about needs and opportunities. New wealth and economic wellbeing stimulates demand for new products and services, increases the capacity for new ideas, and expands the problem set for which new solutions are needed.

Chapter 5

CYCLE VS CYCLE

The Productivity Cycle feeds the Innovation Cycle, and the Innovation Cycle drives the Productivity Cycle (Exhibit 5.1). Innovation is required to prevent the productivity improvement from grinding to a halt due to the law of diminishing returns. Productivity improvement is the net result of commercially successful innovation.

EXHIBIT 5.1 INNOVATION AND PRODUCTIVITY CYCLE RELATIONSHIPS

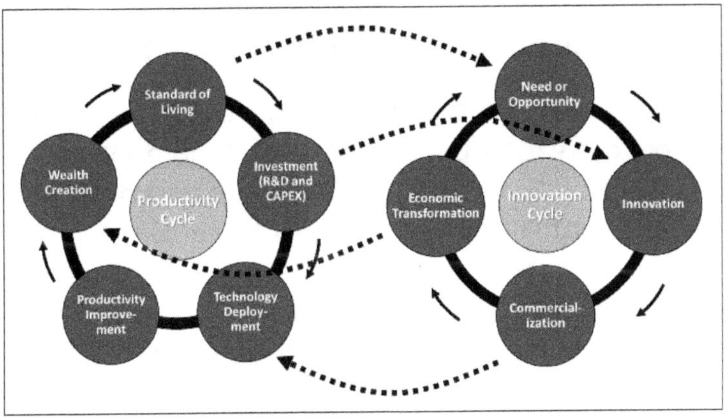

Living Standard Stimulates New Needs

Whatever the standard of living of a population might be, Maslow's model implies that there will be a list of needs or wants that have yet to be fulfilled. A population can be defined by any number of geographic, legal, social, or cultural boundaries which leads to the possibility that there can be many different lists. It is the pursuit of the fulfillment of these needs that creates opportunities for and demands innovation. The young, growing, and prospering population of the US in the 1950's created needs for innovation in home ownership, transportation, entertainment, education, healthcare. The aging but wealthier population of the US in the 2010's created needs in retirement living, leisure activities, continuing education, health care, and elder care. Where hula hoops might have been considered an innovation that fulfilled a need for mindless entertainment in the 1950's, social networking is creating opportunities for new services and products in the 2010's.

Global demographic trends such as the aging of the populations of most of the developed countries is creating needs for more cost effective and more available healthcare and assisted living. The rapid digitization of economies worldwide is creating needs for new types of education and job training. Trends in the growing use of mobile devices for online shopping and streaming audio/video content is creating opportunities for innovation in retail, news media, entertainment, and recreational industries.

Investment in R&D and CAPEX Fuels Innovation

Innovation requires investment. Imagining an idea of how to solve a problem or to create a product or service more efficiently is not enough. There has to be funding to pay for the discovery and design of new solutions that are good enough to be commercialized and deployed. Investment comes in the form of spending for research and development and or in the form of expenditures on new capital to move new technology into commercialization stages. The sources of investment funding include government agencies, business enterprises, and financial institutions. The creators of innovation include government laboratories, academic research laboratories, corporate research/engineering/development organizations, and individual entrepreneurs.

As an example, the original investment in the technology that led to the Internet came from the US Department of Defense and its Advanced Research Projects Agency in the 1970's. At the time, the primary form of communications between military and government organizations around the world was typed messages over analog telephone lines that were connected through electromechanical relay switching machines. The concept of store-and-forward packet switching was developed to take advantage of the growing cost effectiveness and capability of digital electronics. The concept consisted of converting typed messages into digital words that could be chopped up into fixed sized packets that could then be sent individually across a network of computers and then reassembled and converted back into typed messages. Once it was proven that this could be done reliably and practically, then further research and commercialization was funded in the following decades by governments, corporations, and venture investors.

Commercialization Needs Deployment Success

If the commercialization of an innovation is successful, then the new value of the innovation has been understood by the market, the market has been given a platform for buying and receiving it, and a supply chain has been created to support its creation and delivery. The market has decided that it wants the innovation thereby creating a demand for it. Deployment of the innovation into the market is initiated by the new demand. If the demand grows quickly, then deployment is accelerated. Deployment becomes sustainable when new wealth is created and the financial returns begin flowing back to the operating company providing the innovation and to the investors that paid for its commercialization.

Thus the deployment of an innovation will not be successful unless its commercialization (business model plus investment in operations, marketing, and supply chain) has been successful. And the commercialization of an innovation will not be complete unless the innovation has been deployed into the market.

An underlying assumption for both of these two models (Innovation Cycle and Productivity Cycle) is that they are operating in some type of market-based economy rather than a government planned economy. Each of the experiments with centralized government planned economies in the 20th century failed in part due to the fact that there were no mechanisms to unleash and reward innovation and productivity improvement. There were no incentives for those inventors and entrepreneurs not working for the central government to prosper if they took the risks to do better.

Economic Transformation Fuels Wealth Creation

When an innovation has been successfully commercialized, there often is a disruption in an established industry or the creation of a new industry. Google disrupted the advertising industry. Facebook created the new industry of social networking. Uber disrupted the taxi industry and created the on-demand economy. Each of these industry disruptions and creations created an economic transformation that resulted in the creation of improved productivity and new wealth.

Typical indicators that an economic transformation has occurred in an industry include a sudden significant improvement in productivity, a rapid creation of new market value of a corporation, the formation of new ventures, or the merger of corporations that had formerly been leaders in the industry. During the first "dot.com" boom, an email company with relatively small revenue and no entertainment content (AOL Inc.) acquired one of the giants of the entertainment and cable industries (Time Warner) because the public market capitalization of AOL was high compared to Time Warner's. Today's example would be a company like Uber Technologies Inc, that is considered a transportation company even though it owns no vehicles nor employs any drivers. Uber has a market valuation larger than Ford or General Motors. While continued improvements in productivity at Ford and GM have helped add a few billion dollars of incremental new value over the last five years, Uber's innovations over the same period have created more than $80B in new market value where there was none before.

The point here is that an economic transformation of an industry leads to and is the direct cause of the creation of new wealth. The new wealth is created because the innovation behind an economic transformation has resulted in better ways of doing things, in doing more with less, or in fulfilling a need that had gone unmet before..

Chapter 6

How Does Technology Improve Productivity?

Technology Alone Is Seldom The Solution

We all live in a world of processes. Whether it be a business process where we work, a personal process where we live, a medical process where we seek care, or a government process which affects our behavior, the set of activities and interactions we engage in forms a process. The application of technology to any process does not guarantee that the process will be better or more productive. If care is not taken to understand how a process benefits from the application of technology, then a successful implementation is unlikely with the end result being counter to what was expected.

One might think that the impact of automation on what might be called tradeable goods (e.g. manufactured items, raw materials, food stuffs) should be straight forward, direct, and linearly applied. However, there are countless examples of what happens when automation is applied to a process without a thorough analysis of what activities are to be automated and how the new automated process will operate within the environment in which it exists.

When an assembly robot is installed to replace a human assembler, one would expect that the productivity improvement would be measurable almost immediately. However, it depends on how effective the robot is, how effective the work cell has been organized, and what new human tasks have been created by the design of the automated work cell.

There are many examples of how poorly a robotic application can be designed. In the early days of robotics in the automotive industry (mid-1980s), a team of Americans from various manufacturing companies (the author was on the team) visited a truck manufacturing plant in Europe. The plant manager was very proud to show everyone his first robot cell. It was a one-armed pedestal assembly robot that was configured to insert and then screw 2 cm diameter bolts into threaded holes on an engine block one at a time.

The scene was comical. The visitors stood in amazement as the robot (it was blind with no vision guidance) would miss 1 out of every 8 threaded holes and the bolt that missed would drop to the floor of the work cell. After a few minutes, there was a large pile of bolts on the floor that a worker (probably the assembly technician that was being replaced by the robot) who was standing by at the cell had to shovel up and discard. The bolts that were not damaged from falling on the cement floor were recycled for another try by the robot. Given that there likely was no labor replaced (a cleanup person had to be stationed at the cell) and that the cost of quality may have actually increased (due to the damaged bolts), the ROI on that installation might have been negative. Of course, after a few months of improvement in the design of the robot system including various sensors to insure an accurate placement and twisting of the bolt and improved fixturing so that more than one bolt could be inserted at a time, the financial benefit from this one cell turned positive.

Fast forward 10 years to the 1990's. At that time many companies decided to invest in enterprise information systems partially in response to the Year 2000 software bug panic and partially due to the hope that productivity would be improved by a massive amount of new information technology.

The potential productivity benefit looked impressive on paper: a company would convert all of the independent and unconnected paper based processes and personal computer based processes that it had implemented over several years into one centralized software system. The benefits promised included reduced time to fulfill orders, higher quality due to fewer mistakes from the manual systems, and more data which could be analyzed for marketing purposes (the term "big data" had not been invented yet).

The change usually was traumatic. Every employee had to be trained to use the new computer interfaces. Quite often the software implemented forced new steps to be added to the original process to satisfy the needs of the forms built into the software. Initially the time to fulfill orders increased rather than decreased. The new user interfaces were based on generic forms built into the software. The new user screens increased the amount of "paging" through forms to complete the full set of details demanded by the system. The cost of the IT department increased by several multiples because of all the new equipment and software engineers required to maintain and update the new enterprise system.

It became part of the implementation plan for many companies to expect low output levels for 1 to 3 months as an enterprise system installation ramped up. This allowed management and employees time to learn how to use the system and how to adjust all of the workflows so that the output of the plant could return to normal. It would usually take 6 to 12 months after an installation before productivity improvement and financial benefits would materialize. The good news was that there were always lessons learned from each installation that made the next installation project run more smoothly and to ramp up more quickly.

Fast forward another couple of decades to the 2010's and the health care industry is attempting to use information technology to improve quality while also lowering cost in doctor's offices and in hospitals. Another comical scene reminiscent of the robot bolt assembler occurred the first time my primary care physician walked into my examination room pushing a table on wheels that had a laptop computer on it. The professional medical practice firm for which he worked had decided to invest in new information technology. The end result was that my physician was now tethered to a laptop on a table that he rolled around with him all day as he saw patients.

With his patients, the physician spent most of his time typing or speaking to (the accuracy of the voice to text software seemed to be somewhere below 50%) his laptop and a little time talking to his patients. There seemed to be no decrease in quality of health care provided by the physician, but it was not clear that productivity was being improved in any measurable way. In fact, it appeared that the physician had more physical tasks (that resembled furniture moving mixed with data entry) and not less.

Despite the several personal experiences with poorly implemented automation highlighted above, there are many more positive experiences that greatly outnumber the negative ones. From all of these experiences, there are several lessons learned that are applicable to future implementations of automation, AI, machine learning, and robotics technology.

Designing The Workflow Is Essential

Productivity improvement does not always happen in a consistent smooth fashion but quite often in fits and starts. Bursts of productivity improvement that last for years (as in the 1990's in the US) will often be followed by long periods of lackluster or nominal growth (as in the 2010's all over the world). The lesson that has been learned repeatedly is that understanding, designing, and optimizing the workflow in a process is the only way to achieve success in attaining the highest levels of productivity, quality, and throughput.

Over the years, the concept and rules for optimizing a workflow has gone under different names and has been advocated by a wide variety of workflow improvement professionals (Beavers, 2000). One of the first and probably most famous terms for it was Taylorism named after the famous time and motion studies expert John Taylor. Taylor helped usher in the second industrial revolution in the early 1900's by performing studies of how workers performed their tasks to determine how much time was spent on each motion used in the task.

For example, consider a bicycle assembly work station where the person assembling a bicycle was timed to spend 10 minutes on true assembly steps (mating one part to other parts), 20 minutes on organizing the parts that were to be picked up during assembly, and 30 minutes on moving around the work station to find the tools needed to adjust, tighten, and attach the parts on the bicycle for a total of 60 minutes. Taylor would experiment with ways to eliminate the time that does not "add value" to the final assembled bicycle.

For instance, he would redesign the workstation so that the parts the worker needed did not need reorganizing (to eliminate as much of the 20 minutes spent on part organization). Then he would redesign the tooling arrangements and possibly design new fixtures to make the tooling within easy reach (to eliminate as much of the 30 minutes spent on finding tools). When finished, the redesigned work flow for the bicycle assembly would be more like 12 minutes (10 minutes on true assembly action, 1 minute for part movement, and 1 minute for tool movement), resulting in an 80% reduction in assembly time (500% productivity improvement).

The productivity improvement is not the only good news. All the activities associated with finding things and moving things that have nothing to do with the actual assembly, i.e. non-value adding activities, have been eliminated. These non-value adding activities are also often the cause of errors so quality has been improved. Another source of good news is that the assembly person does not have to spend the majority of time looking for things in the work cell, is less prone to make mistakes and likely more satisfied that the result of his efforts are contributing to the creation of a finished product rather than housekeeping in the work cell. In fact, the assembly person often has the best information on what activities should be eliminated and how the work site could be improved.

The body of knowledge that started out as Taylorism has increased and now goes under a variety of names such as Total Quality Management (TQM), Business Process Reengineering (BPR), Just-in-Time (JIT), Toyota Production System (TPS), Design for Manufacturability (DFM), Joku-Riku, etc. These names have all been created to capture the essence of the same basic concept: Design the workflow to minimize non-value adding, problem creating activities.

Improving workflows that are manually intensive (e.g. where over 50% of the activities in the workflow are performed by humans) usually does not require much more analysis than the time and motion studies typified by Taylorism. Time the person and then find a way to eliminate the motions that do not add value. If the initial workflow was designed without any consideration of motion optimization, there should be much low hanging fruit of non-value adding activity to harvest. Eliminating the wasted motion of human workers by providing them with better tooling and fixtures (all CAPEX investments) has been a reliable source of productivity improvement in every manufacturing industry over the decades.

The next level of improvement in process design is to automate those activities in the streamlined process where it is practical, cost effective, and fits within the goals of the process. The original innovation of Henry Ford was to create a track or conveyor belt that automated the movement of an automobile body as it was being built up to a finished vehicle. This assembly line concept improved throughput and reduced quality problems by mixing the most practical automation at the time (convey belt technology) with workplace design and the skilled assembly workers.

Today, an automobile assembly plant is still based on automation moving vehicles and parts to work cells, individual robots that perform assembly and testing tasks, as well as people that are working with the automation. The tasks that people in the factory perform include some old ones and some that are new. The old tasks that are traditional include certain assembly steps for which it is too expensive to fully automate yet. A few decades ago, the instruments on the dashboard of a car were assembled individually. Now the dashboard is an integrated assembly built on its own production line which one worker can now install with the assistance of a robot in a few seconds. The new tasks for people include monitoring systems control centers, maintenance of the electronic and mechanical automation systems, and data analysis of how to improve the existing business processes.

Another lesson learned is that the design of a product should take into account manufacturability and service process steps as well as the required product functions. Usually the best result comes from the people improving, streamlining, and automating a manufacturing process working with and sharing knowledge with the engineers designing the product. This collaboration usually leads to the creation of a knowledge base of best practices, process details, and materials properties that benefits from the use of automation and AI technology. People get smarter about processes and machines get smarter about performance.

The decades long history of automating physical activities in processes has evolved to the automation of information intensive tasks. The operation of an automobile production line is now driven by a complex information system that combines purchase orders, bills of material, machine status data, employee data, and facility data. In reality, the automobile being built just happens to be moving through a big computer center.

Applying the same productivity improving automation of manufacturing process lessons learned to service processes has been making progress over the last couple of decades as well. However, the challenges to automate service processes seem to be bigger in general than for physical processes. Some of the challenges stem from defining and designing the processes. Some of the challenges stem from the organizations, traditions, policies, and even legalities that own or control the processes. That is one of the reasons that industries with physical processes like agriculture and automobile production have seen major improvements in productivity over the years and the medical and educational industries have not seen nearly as much. More on this later.

Improving services processes should start with design which is then followed by technology implementation and training of people. A physical process such as in manufacturing usually means taking some parts and materials and producing something that can be then be shipped to a customer. A service process such as in a restaurant means an ambience has to be created. It means that customers that appear randomly have be instructed on what is on the menu, have to have special needs or desires considered (hold the pickles), and may have need of special treatment because of attitude issues. It means that food has to be prepared using perishable materials, presented in a pleasing format, and then served while at the desired temperature. Then payment has to be processed, the table has to be cleaned and reset, and the next customer cheerfully greeted.

Defining the activities in restaurant processes means one thing. Defining which activities add value is something else. Then identifying which activities should be automated is even a bigger challenge.

Defining value in a restaurant depends on a variety of things. Is value found in the making of the cheapest hamburger possible? Is it in having the finest selection of wines? Is it in serving the tastiest seafood? What does tasty mean? Is it in having the check handled and processed electronically? The process questions form a very long list. Designing the workflow is essential to the automation of and productivity improvement of service processes. Knowledge about how to design workflows for service industries is at a very early stage of formation. The need for service industry improvement is essential, the need for innovation is compelling.

Eliminating Variation Is Essential

Another key element in the improvement of productivity is the improvement of quality. Improving quality means eliminating errors, mistakes, faulty materials, and faulty workmanship. If a production process is perfect, it results in the same part, made the same way, with the same materials, and with the same work steps. It means the elimination of variations in the way an activity is performed, the way a piece of material is prepared, and the way a part is assembled. The American statistician, Richard Deming applied the necessary mathematical principles to how a workflow should be measured beyond the time and motion studies of Taylor. Deming was hired by the Japanese government in the 1960's to help them understand how to measure and improve their manufacturing processes so that they could become more competitive in the global market place that the country was facing after it had struggled to recover from the aftermath of World War II. In the 1950's, Japanese made products (primarily consumer electronics) were laughed at by American customers for being cheap and poorly made. However, beginning in the 1960's, the Japanese industries searched for the best practices around the world to help them improve their global competitiveness in quality and cost (i.e. productivity). They found many of the answers in the teachings of Professor Deming. Beginning in the mid 1960's, products made in Japanese factories became global benchmarks for quality. It was not until the late 1980's after Japanese based electronics and automobile companies had won considerable global market share that US based companies began to study and adopt the lessons that Deming had taught decades before.

What Deming taught, and what the Japanese perfected in many of their industries, was the concept of minimizing the variation or variance in each manufacturing activity.

A simple example is bolt tightening. Assume that the design requirement for the torque on a tightened bolt on the manifold of an automobile engine is 49 lb-ft to 51 lb-ft in order for the engine to operate with maximum power and lifetime. Human operators with just a simple wrench might tighten the bolt to a torque level that varied by several lb-ft depending on how tired they were, how much coffee they had recently, or how much weight training they had.

The chart in Exhibit 6.1 demonstrates the statistical impact of the human with a simple wrench assuming that the variation of the human's wrenching actions occurred according to a normal distribution (typical for many natural processes).

If the goal was to have 100% of the bolts torqued within the range of 49 to 51 lb-ft, then the standard deviation of the wrenching process has to be 0.333 lb-ft. If the human operator had a standard deviation of something more human-like of 5 lb-ft, then 86% of the bolts will have been tightened outside of the acceptable ranges.

Statistical process control analysis of this one action would dictate that bolt tightening would have to have automated assistance to assure that variation outside of the acceptable range is minimized.

EXHIBIT 6.1 MANUAL TORQUE WRENCH DISTRIBUTION PLOT

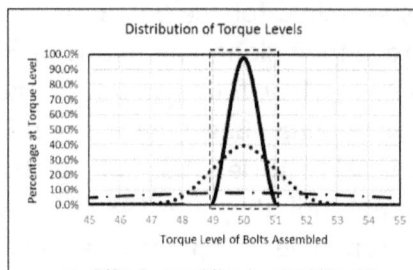
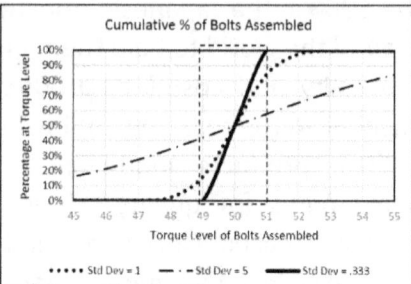

If the bolt tightening activity is to be fully automated with the application of a bolt tightening robot, then the robot must have sensors and actuators to know when it has a bolt ready to insert, find the threaded hole for insertion, rotate the bolt the appropriate number of times to achieve proper tightening, and apply enough pressure on the rotated bolt to achieve the desired tightening torque. This would all have to be done under computer control which has algorithms written to perform each of the physical actions and then report back to a process control algorithm with all the measurements taken by the robot sensors.

Including Flexibility Is A Virtue

One could draw the conclusion from the discussion on eliminating variation and from the assembly line case example that the only way to design a workflow for anything is for it to be rigidly structured so that only one type of product of service can be delivered by one workflow system. However, even in the automobile industry which was the first to use automated assembly lines to produce high volumes of complex products, it became clear that flexibility in assembly was important for satisfying market needs as well as for productivity.

Despite Henry Ford's early marketing statement that you could buy a Model-T in any color as long as it was black, the auto industry learned from competitive pressures that customers would make decisions on the basis of colors, features, and styling differences over sheer price and basic functionality. In order to provide customers with a variety of styles, colors, and accessory features, the product (the automobile) and the workflow (production line) had to be designed to allow flexibility in assembly. So while each part had to be manufactured with the minimum of variation in its workflow, the final assembly process had to be designed to allow for the maximum flexibility in assembly configurations all while maintaining control on cost.

Over the years as the power of computer technology was applied to ever more detailed activities, every step of the production process could be controlled by and synchronized with a plant-wide central production schedule and full data base of bills of materials. The result is that in most auto plants today, every chassis coming down the line is tied to a specific customer purchase order or product marketing planning order. Each vehicle can and will have a different combination of colors, upholstery, accessories, and body features. No longer do companies build blue pick-ups on Tuesday and yellow convertibles on Thursdays. The industry has evolved to a "build-to-order" system because the best practices in supply chain management, product and production line design, and the power of the automation hardware and software have made it just as cost effective to produce low volumes with high variety as to produce high volumes with low variety.

The lessons from this evolution in workflow automation is that quality has to be designed in, low cost has to be designed in, and flexibility has to be designed in. As the spectrum of automation tools and technology available continues to grow, automation of flexible workflows continues to expand. And with it comes the growth in the establishment of smaller, more efficient, more decentralized manufacturing facilities across a wider array of industries. As manufacturing activities become more automated and become more easily configured into smaller decentralized facilities, the location of those facilities will depend less on low wage rate labor and more on location of highly skilled labor and high value markets.

As this trend continues, the ability to automate service industry workflows will expand as well. Service industries depend on flexibility, data, mobility, and customer interaction. Whether providing people with food, entertainment, transportation, healthcare, education, insurance, banking, or protection, service industries involve the collection of data and conversion of data into useful information or actions. The process lessons learned in manufacturing industries all apply to service industries. Each successful innovation in the use of technology to automates and learn from process activities creates new opportunities for future innovations.

Big Data Is Not Necessarily Big Information

Big Data is one of my favorite buzzwords because it could be construed in a lighter moment to mean either a large volume of data or data written with very large font. Putting aside the attempt at humor, the reality is that the digital world is generating enormous amounts of data which is not yet generating very much value.

A 2016 survey (Veritas Technologies LLC, 2016) of thousands of people from more than 20 countries indicated that 52% of all information currently stored and processed by organizations around the world is considered 'dark' (data whose value is unknown) and another 33% of data is considered redundant, obsolete, or trivial (ROT) for a total of 85% of all stored data. This represents trillions of dollars of spent annually to store and protect data that is not creating value.

There are several reasons for the accumulation of useless or redundant data. One reason is there is no clear personal or organizational policy on why data is being collected and when it should be saved or archived. A second is the impression created by cloud computing services, free email storage, and consumer advertising that data storage is essential free. A third is fear to delete data driven by data hoarding tendencies or potential legal liability. A fourth is the expectation that the new value in the data will be discovered at some point in the future.

Now that everyone (people and organization) is living in a digital world where data is rapidly piling up, the questions of when to save, sort, recycle, or trash data are becoming just as important as what is to do with physical goods. The answers to these data storage questions are not obvious because of the new attention being paid to big data by a new generation of data scientists. One of the reasons the term big data is popular is that it helps highlight the promise of benefits from the use of statistical methods applied to processes that have not traditionally been subjected to intensive data analysis.

One of the problems that crops up in the statistical analysis of complex systems in an environment of seemingly large data sets is the "curse of dimensionality". This term was coined by the famous applied mathematician Richard Bellman (Bellman, 1961) who demonstrated that the amount of data required to analyze a problem increases exponentially as the number of dimensions (e.g. variables or states) used in the problem increases. This is especially true when using sampled data for a physical system in the four dimensional world in which we live (length, width, height, and time).

As an example, consider a traffic engineering problem. A traffic pattern analyst has determined that measuring the speed of one car (1 dimension) along 1000 points (10^3 data samples) of the centerline of a roadway between two stop signs was statistically sufficient to give useful information about the safety of the road. However, if the analyst decided that information about air temperature, humidity, rain, surface type, tire condition, car weight, driver age, driver sobriety, time of day, and time of year (i.e. another 10 dimensions) was necessary to answer all the questions about the safety of that roadway, then the number of data samples required zooms to 10^{30} data points (about a million times the size of the total storage capacity of the total Internet in 2014) . At some point in the expansion of the problem solution to include more dimensions of data, the challenge of collecting the data and then processing the data even with the fastest computers becomes either too expensive or impractical. What seemed like a lot of data initially (10^2 data samples) suddenly becomes insufficient by 30 orders of magnitude. The end result for this particular example is that the analyst has to significantly reduce the scope of the problem being studied and to use a much smaller set of data. The big data was not big enough and the information derived from the data that could be sampled might be much less than needed.

One of the more popular areas where big data analytics has successfully taken hold is in professional sports. It originally began when the general managers of sports teams were searching for ways to evaluate new athletic talent to determine who to hire and at what price. Michael Lewis wrote a popular book (Lewis M. , 2003) about how Billy Beane, the general manager of the Oakland Athletics Major League Baseball Team in the 1990s, used statistical analyses of the full range of lower and higher salaried players to determine their effectiveness in helping win baseball games. As most of us in the category of armchair experts always knew, the players with the outsized contracts were almost never producing outsized results when compared to equally or even lesser talented players that were younger and cheaper. Again it was all about searching for the highest baseball player productivity. This type of data analytics has since spread to virtually all of the professional sports leagues around the world, mostly with debatable results.

The data analytics of professional athletics has reached the mainstream consumer because of its use now by sports betting companies. In the old "analog" days of sports betting, a bookie or Las Vegas casino would set the odds on a sporting event (primarily to ensure that the amount of money bet on both possible outcomes would be similar), and all the amateur and addicted gambler would make their bets and hope for the best. Over the last 4 years, there have been several new venture companies that have given a new gambling platform to large numbers of sports fans of all ages and demographics that never before placed wagers on the outcomes of games. These firms have created statistical analytical tools and have provided access to the newly digitized and gathered data for various sports leagues, and have created parameters on which fans can bet that go beyond the final win/loss outcome of a game. Now anyone with a credit card and a mobile phone or laptop can bet on win/loss, over/under, performance statistics of individual players, groups of players from different teams, and almost anything else that can be measured.

The message here is that while more data is often useful in maximizing the probability that a conclusion or forecast of a particular event will be accurate, there is usually a diminishing point of value or return on how much data is useful. The data analysis company 586 purchased by ESPN to establish a position in the fantasy game market estimated ("predicted") that the probability of the Carolina Panthers winning the 2016 Super Bowl was 65%. Of course, Carolina lost handily to the Denver Broncos. Looking at the question from the statistics perspective says that if a 100 games were played by the two teams in a 100 days, Carolina might win 65, but Denver would win 35. The obvious challenge for gamblers is deciding which day is it that the one game will be played. Unfortunately there was not enough data to help with that calculation.

As the capture and availability of data increases exponentially with the advent of the Internet of Things (IoT), there is the question of what to do with all this data and how much of it is of value. We can put more sensors in more locations to create more "data", but are we generating more useful "information". Will the variance of the estimate of the yield of a field of corn improve if we have a sensor on every corn stalk or yet on every kernel of corn? Will the uncertainty of physical processes diminish as we gather more data about every element of the process? If we could put a sensor on every molecule in a kettle of boiling water, would we have more information about the boiling time of eggs than if we just dropped a thermometer egg into it? Where is the boundary at which makes further data collection a waste?

A highly publicized use of big data analytics currently is in the detection of customer preferences or market trends. This type of analysis may have value for companies that want to get more productivity out of their marketing and sales activities or in better focusing their product or service development activities. While this might not be considered big information, the analysis does have value and does improve the efficiency and effectiveness of corporate business processes.

A less publicized and less developed use of big data analytics is in medical research. Before a new pharmaceutical or medical device can be used today in general medical practice in the US, it has to be approved by the Federal Drug Administration (FDA). This usually requires years of test data from human trials that are rigidly structured (double-blind, placebo-controlled, randomized clinical trials) with targeted and simplified hypotheses. Does Drug A improve the situation of a patient with Disease B? Does Drug A perform better than a placebo (i.e. a sugar pill or fake drug)?

The results of these tests can have dual meanings. If Drug A only improves the situation of 30% of the patients tested, then the drug is considered a failure. However, for those patients in the 30% group, it could be considered a success. New questions arise about whether there is something common to the 30% group that makes the drug effective?

Quite often, the placebo group does better than the group getting Drug A. In fact, the placebo effect has been researched extensively (Silberman, 2009) with results indicating there are several neurological factors involved in the delivery of any medical treatment that need to be considered when designing and testing new drugs or medical devices. One of the implications of the findings from the research into the placebo effect is that there needs to be more statistical analysis of many data sources (i.e. big data) to determine how to design the drugs or medical devices to begin with and then how to test the effectiveness of the drugs or medical devices.

With the evolution of new thinking about medicine such as personalized medicine, gene therapy, and immunotherapy, there is a growing need for new types of data and deep statistical analysis based on many new parameters that are not always considered or captured during traditional triple blind patient trials.

Personalized medicine is about customizing a treatment to fit each individual patient by matching proven clinical relationships between the medicine and the genetic structure specific to that patient. The theory is that if enough data is collected from experiments, trials, and biological models, then a system of precise diagnostic tests and targeted therapies can be personalized for each patient for use by healthcare providers. It is not clear yet whether the development of personalized medicine will be end up being a true innovation in healthcare, but it is clear that big data analysis will be an important part of its development.

Gene therapy is an experimental technique that consists of inserting modified genes into a patient's cells instead of using drugs or surgery. Research into this type of medical treatment is at a very early stage, but analysis of data from laboratory experiments, biologic models, and clinical trials will be an essential part of it.

Immunotherapy is a form of treatment that stimulates or modifies a person's immune system to fight diseases such as cancer. Research in this area will also depend on analysis of data from laboratory experiments, biologic models, and clinical trials.

Successful innovations in these new medical developments would truly herald big information from big data analysis. These research areas are ripe for more AI that can help is in determining which data to collect and how much analysis can be done.

Decentralization Is The Future

Natural systems tend to be very large organizations of very small, naturally selected, self-sustaining units. The human body is composed of organs, organs are composed of blood vessels, nerves, and tissues, which are all composed of cell of various sizes and functions. The cells compete with other cells in their environment, they replicate, perform their function to survive, and then die. Cells are composed of even smaller organisms that control how the cell function and how it evolves over time.

There has been a theory developed about the controllability, stability and survivability of very large systems that are composed of very small independently operating subsystems or units. This theory was once called Chaos Theory because it was originally developed in an attempt to explain why catastrophic events would occur for large previously stable systems in an apparent random fashion. One of the core elements of the theory is the hypothesis that a large system can be composed of a combination of independently operating units. These units are not coupled to a central control mechanism but are operating on their own to survive in the environment local to them within the system. There can be peer-to-peer communications of sorts between the units in the system. It is this systems concept of networked but decentralization operation to which many of our social and economic systems seem to be evolving.

In the early waves of automation, bigger was almost always better for achieving productivity improvement. Every industry had a level of concentration of production where the economies of scale would produce the best product at the lowest price. The biggest steel plant, the largest oil tanker, the largest coal plant, the largest farm, the largest hospital, the largest computer maker would always be able to assemble the greatest amount of capital investment and operate with the lowest amount of overhead per output. In other words, the most productive use of assets.

There is an emerging trend to more decentralized (Exhibit 6.2) yet productive economic units due in great part to emerging new automation technology that is making smaller units of assets productive. Decentralized energy (e.g. roof top solar + battery storage), decentralized entertainment (e.g. self-publishing of books, videos, music), decentralized banking (e.g. bit coins and block chains), decentralized food production (e.g. farm-to-table restaurants), and decentralized manufacturing (robotic assembly cells) are examples of the emerging decentralized economy.

EXHIBIT 6.2 CENTRALIZED VS DECENTRALIZED STRUCTURES

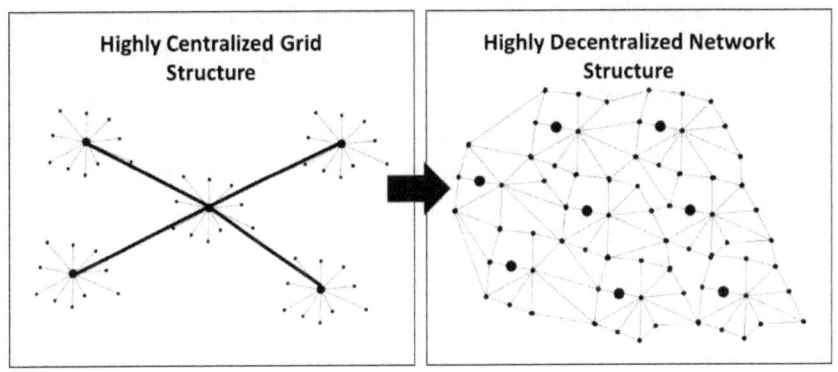

Automation and AI are making this trend to decentralization more practical. As the cost of robotics, processing power, and memory comes down, as the availability of more data becomes available, and as the cost of AI based knowledge and knowhow comes down, the asset level (i.e. CAPEX) at which an economy of scale can exist comes down with it. The on-demand economy as evidenced by the likes of Uber and Lift illustrate how the smallest of economic units can operate albeit within a very large software ecosystem. While it took the genius of the Uber founders to see how one industry (taxi) could be disrupted by providing a software platform that created a highly competitive market that only the purist of macroeconomists could envision, the dispatching, pricing, and resourcing software they created has set the stage for the disruption of and creation of other industry segments.

Automation and AI technology driven decentralization of so many industries can make it less attractive economically for jobs to be exported to lower labor rate countries and more attractive for the creation of new smaller enterprises that serve local markets. The length and transportation costs of supply chains can be shrunken. The needs of a small geographic market can be satisfied by locally created goods and services. Entire generations of technology can be leapfrogged in the less developed economies by implementing more decentralized networks of goods and services rather than depending solely on the building of large grids of infrastructure. Cellular industries have bloomed in portions of Africa and Asia without national grids of landlines. Decentralized generation of power using rooftop solar is gaining market share in developed economies as well as creating new industries in less-developed countries. The entertainment industry in the US has evolved from consisting of a few large corporations based in New York and Hollywood to including thousands of new small producers of media content due to the ubiquity of the Internet and the power of modern smart phones.

SECTION III

INTELLIGENCE: Masters Of AI By Design

Chapter 7

WHAT DOES IT MEAN TO BE A MACHINE?

Do Machines Think or Perform?

There is really no mystery here. Machines are invented and built by humans primarily for the purpose of making the human lot in life better. One might say that human intelligence is creating machine intelligence in its own image to perform human jobs more efficiently. When machines are designed, engineers try not to build in human frailties. There is not a big market for machines that wake up in the morning and then has to decide if it feels well enough to do its job that day although some owners of well used cars may feel like they such a machine.

The human species began to climb to the top of the evolution ladder when it began to realize that a few simple tools might make survival more likely, eliminate the length of hunger periods between meals, or just make it easier to sleep in the cave at night. Over the millennia, the simple tools became simple machines harnessing mechanical leverage to boost human power, simple machines became fuel powered machines that replaced human power, and eventually powered machines became intelligent.

Saying that human intelligence has created machines and artificial intelligence in its own image needs some clarification. Human understanding of self has been evolving over time. In so many ways the machine reflection of that self-image has evolved with it. When people were primarily hunters, they made weapons that helped kill prey for food. When people became farmers, they made machinery that increased the productivity of their muscle power so that they could work the land more productively. When people became industrialists, they made machinery that could increase the productivity of repetitious physical activity. When information started to have as much or more value as commodities, people made machines that improved the productivity of collecting, storing, analyzing, and reporting data.

The image of self that humans have been able to replicate through their machines has always been a fuzzy, out-of-focus, partial image. We only have a rudimentary understanding of how the human brain works, of how the human physiology operates, and what really causes evolutionary change. The machines that we build are just rough mechanical (and now digital) approximations of what little we do know about ourselves and our world.

Alan Turing is considered by many to be the founder of computer science and artificial intelligence (Hodges, 1983). In his 1936 landmark paper, "On Computable Numbers, with an Application to the Entscheidungsproblem" (Turing, 1937), Turing proved that that there was no solution to German mathematician David Hilbert's "decision problem" by describing a simple hypothetical device called a "universal computing machine". In this one paper, he described the initial architecture for binary logic digital computers, operating systems, application software, and memory storage systems. He was also the father of artificial intelligence because he hypothesized that since he had laid out the design of a computing machine, the machine could be programmed to analyze data, make decisions, and eventually think for itself. Prior to this point, the term computer was used to describe a person that operated a calculating machine.

Turing's description of a machine computer led him to later describe an intelligent machine by answering the operational question of "Can machines do what humans do?" rather than on the cognitive question of "Can machines think?" This became the basis of the famous Turing Test (Turing, 1950). An intelligent machine passes the Turing Test if a human evaluator of two subjects, one an intelligent machine and the second being a human, cannot tell the difference in natural language conversations between the two. Turing's focus on the operational performance of machine intelligence reflects the view that people build machines in their own image to perform functions and communicate in ways that humans do.

There are really no mysteries about how a machine thinks, because a machine is designed to think by the engineers that created it. A machine has a set of programs and algorithms that tell it what to do for every potential situation that the machine might encounter. If the designer of the machine code did not conceive of every possible situation that the machine might encounter, then the machine switches into a default operating mode or just stops working and sends out a failure error. Anyone who has ever owned or operated a smart phone, tablet, laptop, or desktop computer has encountered an error message that says the microprocessor inside the device does not know what to do anymore and it gives up. Then there is the universal cure: Hit the REBOOT button! Go back to an initial condition that the machine understands and start all over again. Hope you saved your data! This single experience which every person has had with their "smart" devices should give pause when the thought of driverless cars and healthcare service robots are proposed as an inevitability for the future.

The fundamental fact is that machines can only think the way that humans program them. Despite all of the promise associated with "Artificial Intelligence", the software underneath the AI is still composed of algorithms created by human programmers. Machines can learn if humans program them to learn using learning algorithms created by humans. Machines can be programmed to create algorithms by itself by programming the machines with algorithm generating algorithms.

As long as a machine is subject to the same data and knowledge to which humans are, and as long as a machine is programmed with algorithms created for it by humans, it will not be "smarter". A machine might be faster at gathering data, analyzing it, and returning an answer, but it will still be just a tool built by humans for humans.

A Few Thought Experiments

Let's assume that we have built the ultimate AI-based super computer system (and name it SolaBrain). Let's give it access to all recorded knowledge. Assume that the computer operates at the highest speed conceivable by current science, that the computer gets data from all sensor sources that exist, and that the computer can communicate with humans in a natural language. Does this mean that SolaBrain is smarter than humans, should dominate humans, or replace humans?

Games

If we want SolaBrain to play chess with a human grand master, SolaBrain might win, or lose, or tie. Theoretically, no one optimal strategy exists for victory. If SolaBrain was programmed to exhaustively calculate all of the possible scenarios of moves after each move made by the human contestant, it would have to compare 10^{120} possible game variations or calculate the next best move for each of 10^{43} possible board positions. Assuming SolaBrain is operating at 1000 petaflops (10^{18} floating point operations which is 10^3 times faster than any processor ever built which is 10^9 faster than the fastest 2016 desktop computer) and that it takes 100 moves (or floating point operations) for each game variation, SolaBrain would take 10^{96} years to calculate all of the variations from which to choose. The universe is less than 10^{10} years old as far as we can tell. So there is not enough time for exhaustive calculation by SolaBrain.

An alternative strategy for SolaBrain would be to reduce and simplify the set of possible moves based on previous games played by all grand masters in history or by having data on the general tendencies of the particular human competitor that SolaBrain is facing. This leads to a strategy for SolaBrain to play the game rather than solving it. SolaBrain would have to be programmed to make a probabilistic decision with uncertain data based on prior analysis of relevant data. This is a reflection of how a human chess player would operate. SolaBrain might be faster to come to a conclusion about a decision, but it will still be operating in the human environment and would have to be programmed to make calculated guesses by taking into account its human competitor's experience and tendencies.

On the other hand, the simpler game Sudoku that many play as a brain teaser is solvable mathematically and computationally. Assuming we assign SolaBrain the task of playing Sudoku, it would take a few days to calculate every possible 9X9 Sudoku game variation and just a few milliseconds to calculate a solution for a Sudoku game that starts with 18 given positions (less than 1 second with a typical 2016 laptop). So SolaBrain can solve a Sudoku puzzle much faster than a human, the human might find it more intellectually rewarding to play the game with a pencil rather than just feeding a game to SolaBrain and waiting for the fast machines answer.

Consider the IBM computer named "Watson" named after one of the founders of IBM. Watson is a software system developed by IBM engineers over a period of years that runs on an IBM data center with thousands of servers. On February 16, 2011, Watson was victorious over two human contestants on the popular TV game show called Jeopardy (Markoff, 2011). Although Watson had something of a home-court advantage (the event was held at an IBM research laboratory in front of an audience of IBM executives), it really did nothing more than convert the questions read to it by the host (voice-to-text and then natural-language-processed) and then zoom through its memory banks to find an answer. Although Watson won the three day tournament, the human contestants did score and win a lot of cash.

And oh, did Watson put humans out of the job of being contestants on Jeopardy. No, after this event, Watson was never to be seen on the game show again. The TV game show Jeopardy has been broadcasting every day over the 6 years since (as of the publication of this book) with human contestants for a human audience. The only conclusion that one might draw is that Watson did nothing to improve the popularity or productivity of the TV show and was therefore not a successful innovation. On the other hand, IBM is continuing to develop and market a commercial version of Watson as a smart assistant for a variety of industries including healthcare and advanced research.

There is another conclusion from these excursions into the world of games. While a computer can be designed to play human games faster, the nature of the game gets more complicated strategy, risk, and probabilistic decision making (all human invented algorithms) become more important than speed. The machine is a tool and needs the human to define the game and how it should be solved.

Medical Diagnosis

In many ways, the practice of medicine is similar to a chess game. When a physician meets with a patient, listens to their complaints about pain, makes observations, has tests performed, and reviews all of the data, decisions about a diagnosis of the malady and a plan for treatment must be made. When all the test and observational data is finally analyzed, the physician uses all of their experience, training, research, and continuing education to make the best diagnosis possible and to prescribe the most promising course of treatment. This is a complicated, variable risk, game played with rules that are constantly changing and some of which may be unknown to the players. Test data has uncertainty and error risks. Observations have a degree of interpretation subjectivity. Patients have sensitivities or allergies that can complicate treatment. Prescribed medicine may have side effects and may not have the same efficacy with each patient. The physician's skills vary depending on the quality of their training and the capacity of their intellect.

Would SolaBrain from our thought experiments do a better job than a physician? SolaBrain would still be using the same uncertain noisy data and working with the same knowledge base of conflicting and incomplete data to which the physician has access. SolaBrain would still have to make probabilistic decisions based on calculated risks. What SolaBrain could do is optimize the chances of success because of its access to data and knowledge that no one physician could have and by applying best medical decision practices that have been validated by the medical community.

Physicians today collaborate with each other on difficult cases or cases that fall outside of any one physician's specialty. Physicians today are also collaborating with online medical information systems for assistance in making the best diagnostic and treatment decisions. Patients are going online to augment medical information that they have received from their physicians or in lieu of visits to their physician. A medical version of SolaBrain would be a very useful tool for the healthcare industry, would become a disruptive factor in the practice of medicine, and drastically change how physicians and health care practitioners would function. A medical SolaBrain would likely improve the quality of healthcare by assuring the best diagnosis and treatment recommendation available for each patient based on all medical knowledge. It would also improve the productivity of the healthcare industry by increasing the number of patients that could be diagnosed and treated per unit of healthcare labor or dollar input.

There have been attempts to build a medical SolaBrain for decades but technology has not yet evolved to be at the Watson-on-Jeopardy level of capability. Machine intelligence should be able to improve medical care quality, accelerate the distribution of best practices throughout the medical profession, provide greater clarity of health care options to physician and patient alike, and improve the productivity of the medical profession. The practice of medicine is many times more complicated than a chess game so it will likely be a task that defies an ultimate computational solution.

One conclusion here is that healthcare providers need the intelligent machine as a diagnostic collaborator and as a care giving assistant.

Investment Decisions

Would SolaBrain help make investment decisions easier and more successful? Would people with the SolaBrain become the richest people? Would SolaBrain itself become rich?

There is a growing number of computer algorithms being used in the financial securities and commodities trading industries.

One example is the growing use of multifactor investor strategies in the Exchange-Traded Funds (ETFs) industry which represents about $3.5 trillion in assets. An ETF is an investment fund that holds assets such as stocks, commodities, and bonds. An ETF is traded on stock exchanges. A multifactor investor strategy has as its goal to combine several different themes into one fund. A list of possible themes that could be combined include: (1) consistent dividend flow, (2) stock price upside, (3) low risk of downside, (4) social responsibility, (5) access to ventures, etc. This type of diversified portfolio would be run by a collection of selection algorithms, i.e. a committee of robots (Burger, 2016)

Another example is in high speed trading (Lewis M. , 2014). High speed trading algorithms are helping some firms create extra profit by moving faster that the time it takes for an average stock transaction.

In the high speed trading application, the speed of computers and the quality of algorithms to analyze trading patterns are used to make trades on computerized trading exchanges. A contest between a human deciding to make trades versus a computerized trading algorithm quite often will be in favor of the machine because the machine can detect trends faster and apply decision logic faster. However, as more machines are deployed by brokerage and investment firms to make trading decisions, the market becomes more like a game played by computer algorithms rather than a financial clearing house for people making decisions about their financial wellbeing.

The problem with the computer trading game is that many of the game playing algorithms will be playing different games with different rules at very high speeds. The risk is that computer trading markets could become unstable if one or more algorithms had destructive tendencies by accident or on purpose. Algorithm caused flash crashes have already happened (May 2010, August 2015) in US trading markets. To put controls on the computer trading markets, computerized trading market owners are deploying algorithms to monitor the impact of trading algorithms much like a Spy vs Spy or SolaBrain vs SolaBrain contest.

A conclusion and emerging trend in the electronic exchanges is that there is a need for new algorithm-inspired rules for trading with new enforceable speed limits on trading so that supply and demand forces are not overcome by speed.

Scientific Discovery

Intelligent machines today are already used in virtually every area of scientific research. Let's assume that we have programed our ultimate computer with algorithms that allow it to make new hypotheses and conduct experiments to test them. Let's also assume that the machine made a discovery about the physical universe or about human nature that was profound or fundamental. Let's assume that the machine did not communicate its findings to its human administrators. Would the machine then start using this new knowledge kept secret from humans in its operations and decision making? If this were the case, then an argument could be made that there could be a divergence between human intelligence and artificial intelligence and humans might not understand how to maintain control over the machine.

A conclusion here is that humans should and will continue to design the machine intelligence to be transparent and make all of its data and knowledge available to its human operators. Humans must design machines to produce social or economic benefits (which includes scientific discoveries) so that the cost of building and operating the machine will make a return on its investment.

Machine Self-Preservation

Most of the original funding for the development of computers was launched during war time (World War II) for the purpose of decrypting enemy messages and for calculating firing solutions for artillery. Investment in computer technology for military purposes has continued and has contributed to many if not most of the major technology breakthroughs that are the fundamental technical elements. Every major weapon system now uses computer technology to enhance its performance, to reduce the cost of a solution, and to reduce the risk of casualties to human war fighters.

What if our ultimate computer was programmed by the same company that was building machines for military applications? And suppose that some of the intelligence coded into the military machine to defend itself from hostile action was included in the "civilian" machine. Would the civilian ultimate computer become belligerent and out of control if it felt threatened? What if the ultimate computer was going to be decommissioned for budgetary purposes? What if the ultimate computer was going to be scrapped for a newer model? What if an autonomous automobile that was designed to perform taxi duty was destined to be scrapped and replaced with an autonomous skateboard? Humans could lose control over a machine that was designed to defend itself.

Of course for a machine to defend itself would require that the machine have control over tools that would be necessary for it defeat perceived threats. Just to prevent external powering down a machine would have to have control over the power off switch (easy to do) and over the power source itself (not so easy). To prevent a physical assault, the machine would need to be housed in a hardened site or be weaponized with anti-assault weapons. To prevent cyberattack, the machine would need to have extensive firewall protection. To prevent being starved of data, the machine would need to have control over its data sources. For all of these self-defense mechanisms exist, they would have to be included with the original design of the machine.

An obvious conclusion here is that humans should not arm intelligent machines with weapons to defend themselves!

Was Sherlock Holmes the First AI Machine?

One example of the digital future that this author sees is embodied in the fictional character of Sherlock Holmes. He solved crimes that seemed unsolvable, faster than the professional crime solvers (the police), and with virtually no assistance from forensic science other than what he concocted on his own. He did not have super powers or the ability to see into the future or to speak with the recently departed. He did have highly developed abilities to deduce conclusions from a wide variety of seemingly innocuous observations that were to him evidence and to others less well trained in the deductive process just everyday things to be ignored. Critical to his deductive process was his awareness of facts or information that could be connected to the everyday things to be ignored. He could deduce from the water droplets on a dead body's coat that the victim must have been visiting the south coast of England during the previous 24 hours because that was the only area within a day's train travel that had received rain. Was this deduction due to a super intelligence or due to the ability to remember or access information that others could not or chose not to? Yes, Sherlock Holmes represents what the future of AI can and will be, a machine that can access significant amounts of data in a speedy fashion that results in a conclusion that has the highest probability of being correct.

Was Sherlock Holmes always correct? No, although the number of times he was incorrect depends on the author or playwright that is trying to add to the legacy created by Sir Arthur Conan Doyle (Doyle allowed Holmes just one mistake). The point is that the amount of data to which a decision maker (or algorithm) has access to determines the probability of correctness of any decision. However, in the absence of perfect information, there will always be some level of probability that a decision maker will be wrong.

A simple conclusion: Sherlock was the first popular AI machine!

Although the thought experiments above do not provide ultimate proof that artificial intelligence is not a threat to dominate human intelligence, they are intended to convey a simple message: Artificial intelligence will dominate human intelligence only if humans surrender to it. Human intelligence is designing artificial intelligence to perform as a tool, as an assistant, or as entertainment. Surrendering to AI would have to be by design as well.

Chapter 8

WHAT DOES IT MEAN TO BE A HUMAN?

Evolution Of Human Intelligence

Many a historian, philosopher, psychologist, biologist, geneticist, and cleric have written about it means to be human. Looking at human DNA does not help very much since 96% is shared with chimpanzees (Lovgren, 2005) and 60% with bananas (National Human Genome Research Institute, 2010). Now that the genomes of humans and other species have been identified and compared, the conclusion of many scientists is that each step of evolution has been a process of adding onto the DNA sequences of previous steps. Much of the common DNA material is even considered DNA-junk left over from distant mutations. And as one branch of species would not survive its environment, another more adapt species would thrive. The human species, homo sapiens, has not only survived but has thrived. It has also had a measurable impact on its environment.

Yuval Noah Harari, an Israeli professor of history and author, hypothesizes ((Harari, 2014) that homo sapiens evolved to dominate the physical world because they were the only animal that could cooperate flexibly in large numbers. This ability to cooperate resulted in the organization of large scale human systems such as religions, governments, trade networks, and legal systems. And he argues that this ability is derived from the fact that humans were the only animal that could think abstractly and believe in culturally unifying non-physical concepts such as gods, nations, money, and human rights.

Looking beyond the high level study of the evolution of humans, David F. Bjorklund, a professor of psychology at Florida Atlantic University, believes that most paleoanthropologist and biologists have focused primarily on "adult" evolution and have overlooked the role of "adolescent" evolution (Bjorkland, 2007). He argues that for adults to survive and evolve, the children they bear must survive childhood to become adults. While it has taken millennia for adult humans to change, adapt, and evolve, there are indications that some form of change in adolescents may be happening in shorter periods.

For example, the age of the onset of puberty worldwide has gotten younger by 2 to 3 years over the last 150 years. But he also argues that while some physical transformations of adolescents are occurring at a younger age, the length of time that it takes for the immature mind of the adolescent to develop has actually been increasing. The period of immature adolescence for humans is longer than any other mammal and has been important in the evolution of humans because it allows a long time for the brain to learn and adapt to its situation. The increase in time that the human mind stays in an immature adaptive learning adolescent stage could be due to increased time spent as a dependent and time enrolled in educational institutions. He believes that the immature adolescent mind is more adaptable to its environmental changes and that there is a benefit to the emerging adult. For example, medical doctors are often in an extended adolescent period until their late 20s due to their educational and training requirements.

And then there is the very curious phenomena that has been called the Flynn effect. The Flynn effect is named for James R. Flynn (Flynn, 2007) who began in the 1980's to publish research indicating that the average IQs of definable populations and nations have been increasing consistently since the 1930's. Investigation of the change in IQ scores has been conducted by many other researchers since Flynn first identified the trend. Many of the researchers were looking for IQ test biases, contributing factors, demographic dependencies, and genetic factors that might explain the Flynn effect.

There have been several factors proposed that might be causing the measurable improvement in IQ. These other factors include improved nutrition, better health care, more years of education, more mental stimulation of children at early ages, and more familiarity with test taking. Although there has been no clear definitive explanation for the upward trend, ongoing testing over the years indicates generally that (1) most of the IQ improvement occurs in the lower half of the bell curve of IQ, (2) populations from less developed countries (lower standard of living) tend to have the lower average IQs but tend to improve the most, and (3) IQ improvement for a population from a developed country (high standard of living) tends to diminish and even stall after many decades of upward change. It is as if there is a limit to how much the standard deviation of the IQ bell curve can be reduced or a limit to how much the "IQ inequality" range can be reduced.

If the collaborative, self-organizing, and abstract thinking traits of homo sapiens helped win the living species evolutionary contest to date, one might ask what role might the evolution of automation play in the future. The collaboration between humans and machines began early on with the creation of the first tools and the wheel. When the first operators of a weaving loom climbed about the marvelous machine, they must have felt a new and unique sense of power provided to them by their mechanical work partner. Today we find ourselves collaborating with automated voice attendants at call centers, smart digital assistants on our cell phones, assembly robots in electronics plants, and ATM machines at kiosks.

We not only collaborate with our technology, we also adapt our behavior because of it. Corporations are organizing themselves on the basis of how the automation they use in their business gets deployed, how it is financed, how it gets designed. People are extending their use of abstract thinking by using technology to do things that they physically could have done before such as social networking with friends and strangers alike across the global geography.

The changes that automation is bringing to our daily lives and workplace is forcing human learning to be expansive, adaptive, and ongoing. Sometimes the changes are subtle, sometimes sudden and disruptive. In addition to the research of Bjorklund, Flynn, and others, there is an ongoing emergence of new insights that indicate there are factors in short term human evolution that are related to the factors in the evolution of technology.

Human As Master Of Machine

As the role and importance of automation and AI grows in economies around the world, one thing that we must remember is that humans are the master of machines, not the reverse. While this sentiment could be viewed as the philosophical equivalent of whistling while strolling by the graveyard, it really is based on the practical and pragmatic aspects of how technology is developed and implemented. There are at least four reasons why humans are now and should remain the master of machines.

The first reason is that humans are the "Masters of Design": design of the mechanicals, software, hardware, system, sensors, user interface, everything. Machines are designed by people for people to serve people. Of course, over the years, the design engineers have developed automation and AI design tools so that they could design more complexity into modern electronics, electromechanical devices, robotics, and all machines for industrial, transportation, healthcare, or consumer applications. Except for the simplest of products, most designs today are completed by teams of engineers and project managers and consist of hundreds or thousands of computer aided drawings and documents. For most complex systems, no one person actually has a full understanding of how a machine is designed; these machines are designed by a collaboration of people, design tools, and knowledge bases. Going forward, human led design collaboration teams will continue to be the master of machine design while ably assisted by design technology tools.

A second reason is that people are the "Masters of Operation": people build, install, adapt, maintain, repair, operate, and update machines. As each wave of new product development brings a greater degree of automation of many of these operational responsibilities (all for the sake of productivity improvement), the human content of machine operations becomes more important and requires more skill. Just as in design, where it now takes a team of people collaborating with smart tools and AI technology, the operation of a machine often requires a team. The more complex the machine, the more skilled the human operational and support teams must be.

A third reason is that people are the "Masters of Ownership". Machines are paid for by people or organizations run by people. They invest in the development, purchase, and implementation of the machines. Machines are purchased for the purpose of creating a financial benefit for the owner of the machine. If machines do not provide a benefit to the owner or to someone the owner wishes to receive the benefits, then the machines will not be built.

A fourth reason is that people are the "Masters of Collaboration". While machine collaboration is a relatively recent phenomenon, it will likely continue to increase across industries and at all levels of automation. Although recent consumer applications of collaboration with machines are the best known, (e.g. Apple's Siri, Amazon's Alexa, Microsoft's Cortana), machine collaboration has a long history starting with computer-based diagnostic and maintenance systems for automobile repair, mainframe computer system repair, and industrial machinery repair. The creation of fully functional voice recognition systems, natural language interpretation, and voice-to-text systems has made collaboration with machine intelligence far effective and useful.

Being human is a dynamic state that changes from adolescence to adulthood, from generation to generation, as well as from millennia to millennia. Science is teaching us that the human species is evolving faster than previously thought and that the permeation of new technology throughout global societies and economies may have something to do with that. We should understand that as we invent and deploy smarter machines that we start out being master and only we can lose that role by surrendering it. We should understand that as we make machines smarter and more productive, we are making the machines evolve to be more "helpful" to us. As a result, we are adapting and evolving in the way we use, interact with, and think about machines. We have always looked at machines being our tools, our extended muscle, out extended cognitive self. We are now finding that the degree of helpfulness we receive from and the degree of dependence on intelligent machines is becoming more collaborative. In reality, we humans and our machines are both adapting and changing.

Chapter 9

MORE RESPONSIBILITIES FOR MACHINES?

At least up to this point in history, the definition of a machine and its purpose for existing has been defined by humans. In fact, that is the point of this entire book and the hypothesis underlying it: that machines are designed, built, and paid for by humans to serve human purposes. For most of our society that has been accepted either implicitly (people who don't think about it and just enjoy the fruits of the machine's output) or explicitly (people who make them, talk about them, or fear them). However, as with everything in today's evolving society, there are some abstract thinkers that are contemplating how to extend or apply human rights and responsibilities to machines.

One example is a report prepared by the Committee on Legal Affairs for the European Parliament (EU Parliament Committee on Legal Affairs, 2016) and released to the public in June of 2016. While the report was only a draft of a recommendation to a parliament that has little real authority in the governance of the nations in the EU, it does reflect the full thinking of a group of lawyers employed by a transnational quasi-governmental bureaucracy. It addressed a full array of possible extension of human rights, responsibilities, and obligations to "self-aware" "robots, bots, androids and other manifestations of artificial intelligence". Of course, the document begins with references to Mary Shelley's Frankenstein's Monster and Prague's Golem among others and cites Isaac Asimov's three fundamental laws for robotics (Asimov, 1942) as a benchmark of rules that capture intrinsically European and humanistic values.

The key takeaway from this draft report is that robots should be "personified" and the result of that personification are entirely new areas of law, government, ownership, and taxation. There is precedent for this. Corporations and partnerships are treated as legal "persons" whose definition have evolved over the last 100 years by the courts and legislation (Mark, 1987). It has taken many years for the criminal and civil limits of corporate personification to reach a level of general uniformity. We are at the very beginning of any trend to personification of robots and whether this trend will continue or remain an academic curiosity remains to be seen.

The legal personification of robots also creates whole new litigation scenarios that could extend the "Twinkie" defense (LII of Cornell University Law School) to the robot. If the robot is charged with a crime, the robot's lawyer could claim that the robot's software had a bit-flip or that there was an power surge in its circuits. Would the robot's creators (software programmers, electronics engineers, mechanical engineers, etc.) be any more liable for a recalcitrant robot than a parent is for raising a serial killer? All this, of course, would be a creator of new jobs for lawyers to defend robots unless of course robots take those jobs as well.

The topics covered in the draft report included: Definition of a smart autonomous robot, liability, ethics, creation of a European agency for robotics, intellectual property rights and flow of data, standardization, safety, security, registration of smart robots, taxation, licensing (designers, users), and extension to International agreements. In context, it is clear that the committee that prepared this draft approached it as a serious task. Amidst some of the bad ideas are things that will likely must be considered. A short discussion of each topic follows.

The report's attempt to define a smart autonomous robot is an example of how an idea with a good intention can be very shortsighted. The definition in the report describes what the authors can read about in the popular press as a machine that acquires autonomy through sensors and/or by exchanging data with its environment (inter-connectivity) and trades and analyses data, is self-learning, and has a physical support. The definition is clear, simple, and could be used to describe a micro-wave oven. The task of creating a definition of a robot is daunting and time-dependent with an expectation that obsolescence of the definition is almost guaranteed. For example, if you empaneled a group of telecommunication experts in 1985 to define a mobile phone, they likely would have described a short wave radio that fit in the trunk of a Buick with a long whip antenna attached to the car and a telephone handset by the driver's seat. It would have been only the most dedicated science fiction advocate on the panel that would have included something like the Dick Tracey wristwatch phone.

The evolution of the definition of a robot person could take the same course as the definition of a corporate legal person has over the years. Today a corporate legal person, for example Proctor and Gamble, is a group of thousands of people, facilities, equipment, and intellectual property. If a fleet of self-driving cars is being owned and operated by one company, should the entire fleet of robot cars be one robot person. The entire fleet is undoubtedly communicating via networks with more than likely some network level of software making decisions about and giving commands to each robot car in the fleet.

On the other hand, clarifying liability for robotics is a quite practical objective. One could argue that the area where robotic liability will first be tested and likely defined after the fact is for self-driving cars. Today many car makers are including self-driving features (parking, speed control, collision avoidance) with more to come. Ultimately, when a fleet of self-driving vehicles does hit the roads, the key liability questions for accidents will include questions of who is at fault which could include (1) the software designer, (2) the computer hardware designer, (3) the sensor makers, (4) the car owner if negligence in maintenance is discovered, (5) the car maker is the car has not been upgraded to the most recent release of software and sensors, (6) the legislators for setting highway speeds inappropriately, (7) the highway designers for not making the roads more self-driving friendly, etc.

Of course if robots become personified and take on the liability for its actions, the obvious next step would be to create insurance policies for robot persons. The whole concept of insurance implies that accidents will happen, someone must pay for the accidents, and that the insured robots are not perfect. However, if robot persons have to buy insurance, how will the robots pay for the insurance, and who will sell the insurance to the robots. Will advertisements be targeted at robots so that they can decide which insurance agency to use? Will fans have to sit through football games watching advertisements for robots policies? Will celebrity robots star in commercials for robots? Maybe this a bit of a digression.

The report suggests another logical extension that derives from the concept of a robot as a person: ownership of intellectual property. If a robot invents a new gadget or discovers a new molecule, the robot will have the right to own the intellectual property and any income that derives from that discovery. This single concept is in many ways the largest leap in the imagination of the lawyers and government employees preparing this report. The leap is that a robot will not only be self-aware, but intellectually curious enough to hypothesize a new theory and then clever enough to devise an experiment to prove or disprove the hypothesis.

One could argue that much of what Thomas Edison did was not so much genius but very much persistence in the search for a practical solution to a problem he had set his mind on. Edison strategy was to conduct as many experiments a possible as quickly as possible to exhaustively eliminate the ideas that would not work. Edison is best known for inventing a cost effective, practical electric lighting device not being the first to invent an electric lamp. It took him 10 years and thousands of experiments with various materials and structures to develop a lamp that had disruptive impact on the lighting industry at that time. Edison was driven by an insatiable curiosity and equally importantly by a financial incentive to generate a cash return from the sale of an innovative new product. And he was also tireless in filing patents to protect his economic return from his discoveries. If Edison was doing this same product research and development today, he likely would be using as many computer tools as possible to perform simulated experiments and to synthesize new material structures using software simulations in addition to whatever physical experiments might be necessary to validate any software findings.

Is there an Edison robot in our future? Or will there be an AI system controlled, guided, and queried by a team of scientists and business entrepreneurs that will be searching for a great new discovery that has economic benefit? When the discovery is made, is it the human team that is using the system that should own the IP? And what does it mean in terms of economic and legal legacy if a machine owns a patent? If the machine is turned off or damaged by a hacker or natural disaster, who inherits the ownership of the patent?

Taxation is never far from the thoughts of government officials and so it is with the personification of robots. If we make the leap and define robots to be persons that can own intellectual property, buy insurance policies, receive income, and have other economic responsibilities, then they can be taxed.

First suggestions include taxing robots for the retirement costs of the people whose jobs were eliminated by the robot. But it would be a short leap to tax any income real or implied generated by the operation of the robot in an economic activity. The unbounded imagination of taxing authorities could even lead to the concept of taxing the extra income produced by the additional productivity improvement created by the use of the robots. Thus if a robot replaces a welder on an automobile assembly line, the robot could be charged social security taxes equal to the wages of the worker that was replace. Additional taxes could be placed on the increased value that the welding robot is producing. If the welding robot is welding at twice the speed of the human (i.e. in reality replacing two human welders), then an additional tax on the extra human worth of output could be considered for a special new tax.

Of course, for all these tax ideas, robots would have to be designed to be self-reporting to the taxing authorities. Or more likely, the companies and individuals that own the robots would be made responsible for reporting the value produced by the robot. Maybe there will be an IRS Form i1040.

Then there is the question of what does it mean for a company or individual to "own" a robot if the robot is now legally personified.

Of course, if robots are going to be personified and taxed, they will need to be registered. Today automobiles are registered and taxed as well as other vehicles such as boats, trailers, and airplanes. Registration of robots, following the concept of the personification of machines, will be more like creating social security numbers for each device. Registration will allow for the tracking, counting, and taxing of the machines.

If robots are going to be registered, tracked, taxed, and sued, should not the privacy rights and rights of legal representation be included? What about the concept of welfare or retirement benefits for robots that have been superseded by more advanced robots or robots that have worn out. This topic will likely be added by the robots themselves if we give them the authorization to expand on this list. Will robot unions be far behind?

In a somewhat contradictory perspective by the EU committee, the personification of robots does not eliminate the ethical responsibilities of the humans that make and use the robots. The committee recommended developing an ethical framework for the design, production and use of robots to complement the legal recommendations of their report. They recommended a framework in the form of a charter consisting of a code of conduct for robotics engineers. They recommended a code of conduct for research ethics committees when reviewing robotics protocols and ethical guidelines for designers and users. They want everyone and every machine to be consistent with "the principles of beneficence, non-maleficence and autonomy, as well as on the principles enshrined in the EU Charter of Fundamental Rights, such as human dignity and human rights, equality, justice and equity, non-discrimination and non-stigmatization, autonomy and individual responsibility, informed consent, privacy and social responsibility, and on existing ethical practices and codes".

And of course no serious piece of work from a government committee would be complete without a recommendation for the creation of a new government agency with new funding and staff to control and monitor the implementation of all its recommendations. The creation of a government agency (European in this case) for robotics would be the least that a serious government could do since it was inventing a new species of legal life called the robot.

A final area covered relates to the more machine-like nature of the robot persons: standardization. Standardization of safety rules and regulations, security policies, data security, and privacy. An easy example of the need for standardization would be the self-driving vehicles. One might not want the highway rules that Denmark might impose on the self-driving software to be applied to the autobahns in Germany where the average speed is 50 km/h higher.

One area not covered by the EU committee's report is robot punishment for illegal acts. No mention was made of what type of robot prison should be created to house robots that have been found guilty of civil or criminal violations of the law.

The bottom line of all of the above hypothetical concepts of personifying a robot is that this would only happen if people gave these rights to the machines that they have designed and built. Humans would have to abdicate their natural role in the evolution of machines and humans. Only in the extreme extension of the definition of humanity would it make sense to extend the rights and responsibilities of humans to machines. It would not be a matter of machines rising up and taking things into their own hands, it would be a matter of humans surrendering their position in society and surrendering their hard won rights to machines that they build and operate.

SECTION IV

NECESSITY: Tech Drives Standard Of Living

Chapter 10

HUMAN NEED DEMANDS MORE TECHNOLOGY

Maslow's hierarchy of needs model is all about that simplest human need: improving the standard of living from where it is to something better. History tells us that improving the standard of living means improving economic productivity. Improving economic productivity means more technology: more automation, more robotics, more artificial intelligence. Looking at the trends of several key factors tells us that the need for productivity-improving technology is becoming ever more crucial.

One key factor is the aging of populations around the world (Exhibit 10.1). The populations of many cities, sub-nation states, and nations (especially those with higher standards of living) are aging because birth rates are declining and expected lifetimes are increasing.

EXHIBIT 10.1 POPULATION AGING FACTORS

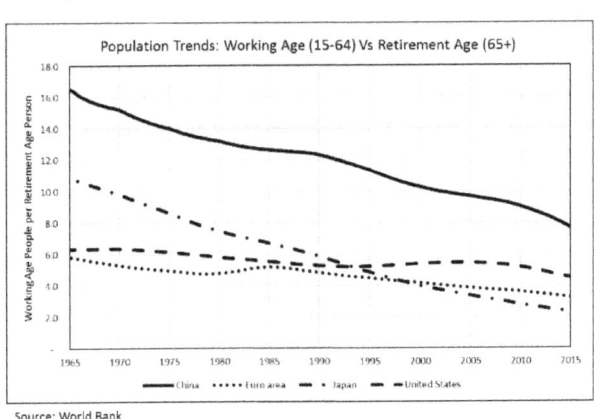

Where this is happening, the proportion of the working age portion is shrinking in comparison to the aging, retired, care-needing portion.

When the social security system was created in the US in 1933, the ratio of working age people (15 to 64) to post-64 age was 18 to 1. In 2015, that ratio was 4.5 to 1. If that trend continues, the ratio in 2035 will be 3.9 to 1. This implies that productivity of the US economy would need to increase at least 3% each year for the next 20 years just to keep up with the aging of the population and inflation assuming that both rates are the same as the last 20 years. The average rate of productivity growth over the last 10 years has been less than half that rate.

In 2016, the National Bureau of Economic Research published a study (Maestas, Mullen, & Powell, 2016) that used data for the US economy from 1980 to 2010 which indicated that for each 10% increase in the share of population above the age of 60. there was a decrease in economic growth of 5.5%. This decrease in economic growth breaks down into a decrease of 3.7% in the productivity growth and 1.7% decrease in labor force growth. These empirical results imply something beyond the need created by the extra cost burden implied by the shrinking of the worker-to-retiree ratio. One answer could be that the education and productivity levels, of each retiree leaving the work force is greater than that of each new young person entering the workforce with the result being that the average productivity level of the total workforce decreases. Another could be that the knowhow and knowledge of the retiree's leaving the workforce is not being captured and provided in training to the new workforce entrants. Either of these reasons calls for technology innovation for use in areas such as business processes, education and training processes, and social networking.

Thus the population aging trend implies a need for new sources of productivity improvement and at increasing rates above what the various economies have experienced in the last two decades. The same is true and even more urgent for Japan, Europe, and China.

Standard of living starting point is another factor. Underdeveloped nations that have a relatively lower level of economic success generally have a need for a more rapid productivity improvement in their economies if they want to catch up to the more economically developed nations. A measure of improving the standard of living in a poor nation might be adding more protein to basic diets or decreasing access time to reach basic health care facilities from days to hours. A measure of improvement in a rich but aging nation might be reducing the cost of assisted living for the elderly. The type of technology required by each will vary according to need, but the productivity improvement available from technology is essential.

The finiteness and renewability of resources is another factor. As the natural resources that are mined, pumped, fished, and harvested are consumed by a global population where standards of living are growing, the value and cost of those resources increases. And as cost increases, the need for productivity improvement in terms of the efficiency with which those resources are gathered, processed, consumed, and discarded. The increasing cost and diminishing inventory of finite resources will increase the value of renewable resources. Renewable resource development and deployment depends on technology innovation even more than that of finite resources. Technology innovations are key drivers in successful economic growth of renewable resources. Innovations in energy technology include photovoltaic semiconductors for solar power and massive wind turbines towers for wind power. Innovations in water technology include desalination plants for sea water conversion, and micro purification machines for water recycling and remote water treatment. Innovations in food production technology include hybridized and genetically modified species to increase yield and quality, agricultural information data analysis to optimize planting and harvesting processes, and sustainable aquaculture process technology that allows the increase of seafood production without further depletion of sea wildlife populations.

Chapter 11

INDUSTRIES THAT NEED MAJOR IMPROVEMENTS

While the bulk of this book is devoted to examining the factors that drive and will eventually limit the extent to which automation will control and replace activities in everyday life, there are several industries where much more productivity improvement is needed as quickly as possible to deal with social and economic issues that have continue to grow unabated. While so much of the productivity improvement that has occurred over time has been achieved in manufacturing and agriculture, there are many other industries that are still in the earliest stages of seeing productivity improvement on similar scales. In the following, an attempt is made to use simple measurement parameters to highlight the relative improvement needs and successes of several key industries.

Agriculture - A Major Success Story

By almost any measure, the US agriculture industry has been a success story for over 75 years. The charts in Exhibits 11.1 through 11.3 are examples of how the productivity of farm labor has increased dramatically and consistently since about 1940.

EXHIBIT 11.1 GRAIN OUTPUT PRODUCTIVITY GROWTH

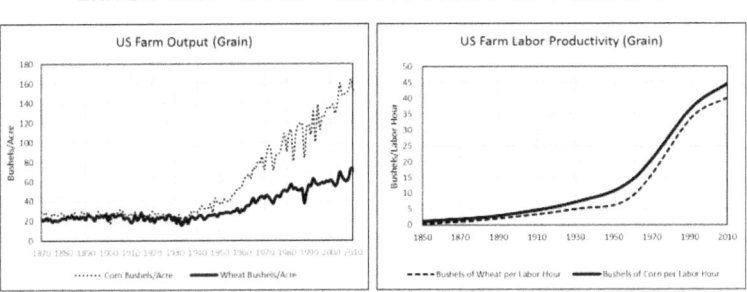

Source: US Department of Agriculture, University of Kansas, University of Iowa

While productivity growth between 1850 and 1930 was small, it began to increase in the late 1930's and early 1940's. During the

1930's the US federal government significantly expanded funding of agriculture research and education as a response to the Great Depression as well as to prevent disastrous farming practices such as those that contributed to the Dust Bowl in the plains states. During the 1940's, the increase in World War II defense spending also contributed to the growth in the application of mechanical technology in the agriculture industries.

During the 1950's, innovations from research in grain hybridization, animal husbandry, fertilizer and insecticide chemicals, and farming equipment all contributed to a continuation of dramatic advancements in food production in the United States.

During the 1960's, much of the technology, chemicals, seed stocks, and agricultural management practices were adopted by much of the underdeveloped world resulting in what has been called the Green Revolution. As a result of the introduction of high-yielding varieties of cereals, such as dwarf wheats and rices, of chemical fertilizers and agri-chemicals, and with innovation in irrigation techniques, agricultural productivity skyrocketed around the world.

American scientist Norman Borlaug is credited with being the "Father of the Green Revolution". He received the Nobel Peace Prize in 1970 for his efforts.

EXHIBIT 11.2 MILK OUTPUT PRODUCTIVITY GROWTH

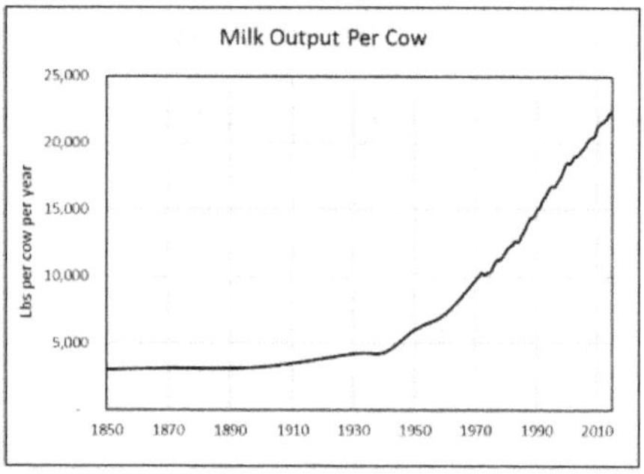

Source: US Department of Agriculture

While one US farmer today feeds the equivalent of 160 people, the US is not self-sufficient in every category of food source. The US is the largest exporter of food in the world followed by Netherlands,

Germany, China, Brazil, France, Spain, Canada, Belgium, India, Italy, Thailand, and Australia. However, the US is also the largest importer of food in the world. An estimated 15% of the U.S. food supply is imported, including 50% of fresh fruits, 20% of fresh vegetables and 90% of seafood.

EXHIBIT 11.3 US FARMER PRODUCTIVITY GROWTH

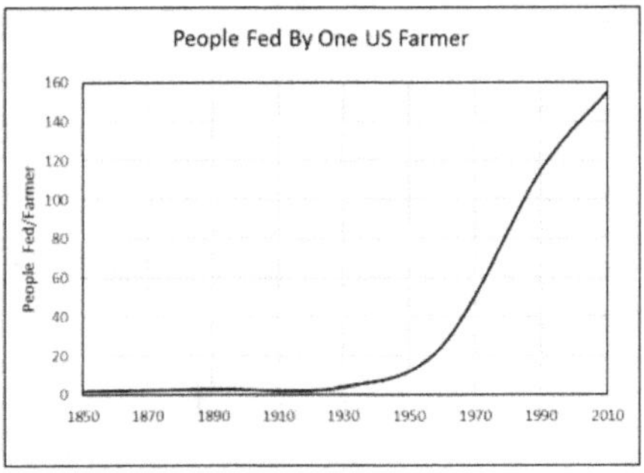

People Fed By One US Farmer

Source: US Department of Agriculture

The seafood industry is an interesting case. The wild capture of seafood globally reached a peak of about 90 million tons in the early 1990's. Since then a global aquaculture industry has developed initially in northern Europe and over the last decade in China and Southeast Asia. Aquaculture (seafood grown in land-based and ocean-based farms) now accounts for about 50% of total global seafood production. China accounts for 62% of global aquaculture production, Southeast Asia accounts for 26%, Europe accounts for 5%, and the Americas account for 4%. The US is the third largest consumer of seafood in the world but is 14th in terms of aquaculture production.

In summary, the US agriculture industry has over the last 75 years has a successful track record of consistently and significantly increasing productivity through technology and knowledge in most food categories, has exported the technology and knowhow behind this to the rest of the world, and continues to be a leader in agricultural research in plant and animal genetics, automation, artificial intelligence, fertilizer chemistries, and preservation sciences. However, one agricultural segment in which the US has fallen behind is seafood production especially in aquaculture. This is a clear market need that is waiting for innovation and investment.

Education - In Need Of Success

The two phrases (1) productivity improvement and (2) education industry are seldom used in the same sentence. If the definition of productivity used in every other industry (i.e. greater output per unit input) is applied to the education industry, the results are disheartening. Data from more than 50 years (Exhibits 11.4) shows that the productivity of the education industry at the elementary and secondary levels has been in a state of decline as measured with a variety of measures.

EXHIBIT 11.4 PRE-COLLEGE EDUCATION INDUSTRY

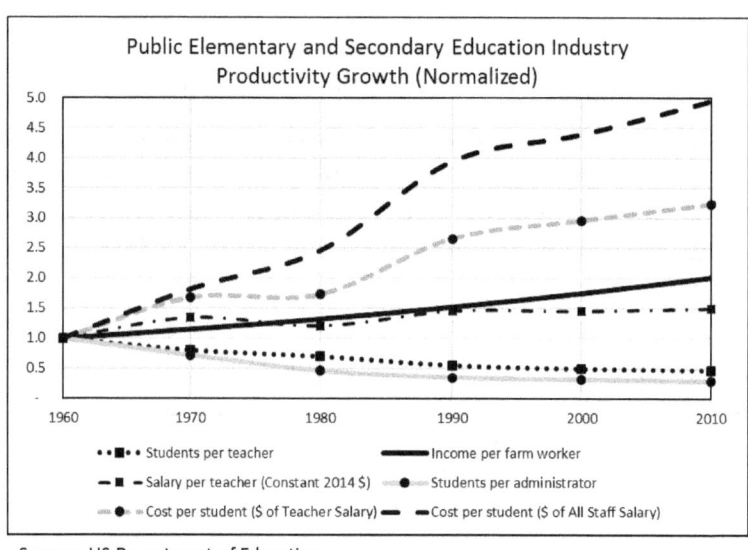

Source: US Department of Education

Today it is taking twice as many teachers and almost 3 times as many administrators to educate one person to the high school diploma level as it did 50 years ago. This is reflected in the increase in the cost of educating each student which has increased by a factor of 5 during this time. The lack of productivity is also reflected in the fact that the salary for a teacher has been flat for two decades and has seen nominal growth over the 50-year period.

The same problem exists in the post-secondary education industry (Exhibit 11.5). While the number of students per faculty or per all staff has been flat for the last 20 years the cost per student in constant dollars has increased more than 50% in the last 20 years and over 140% over the last 35 years.

EXHIBIT 11.5 POST-SECONDARY EDUCATION INDUSTRY

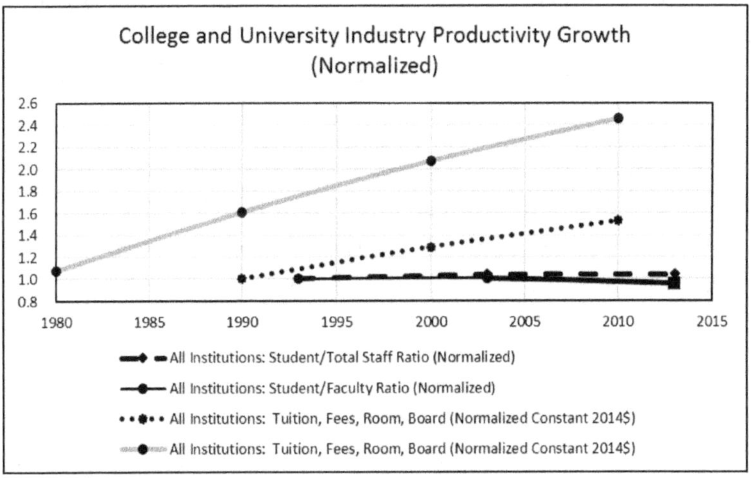
College and University Industry Productivity Growth
(Normalized)
2.6
2.4
2.2
2.0
1.8
1.6
1.4
1.2
1.0
0.8
1980
1985
1990
1995
2000
2005
2010
2015
All Institutions: Student/Total Staff Ratio (Normalized)
All Institutions: Student/Faculty Ratio (Normalized)
All Institutions: Tuition, Fees, Room, Board (Normalized Constant 2014$)
All Institutions: Tuition, Fees, Room, Board (Normalized Constant 2014$)

Source: US Department of Education

Comparing the performance of the US primary and secondary educational system to that of other countries reveals a system that might be considered adequate at best. The Programme for International Student Assessment (PISA) test is administered to students every three years to a wide variety of developed and developing countries. PISA measures reading ability, math and science literacy and other key skills among 15-year-olds. PISA test results from 2015 ranks the U.S. to be 38th out of 71 countries in math and 24th in science. Among the 35 members of the Organization for Economic Cooperation and Development, the U.S. ranked 30th in math and 19th in science. Singapore, Japan, and Taiwan ranked in the top 5 for each category. Most of the European countries, Canada, and Australia ranked above the US in each category.

Looking at the post-secondary educational system of colleges and universities, the top US universities rank high compared to the rest of the world. A 2015 ranking of the top universities around the world (Shanghai Jiao Tong University, 2015) gives the US 18 of the top 25 universities and 33 of the top 50. However, an international test aimed at assessing adult educational skills developed by the Program for the International Assessment of Adult Competencies indicates that the average college or university in the US produces degree holders that rank about the same as the test scores from the US secondary schools. In the 2012 test results, Americans below the age of 29 with a bachelor's degree or better ranked 16th out of 24 nations (Carey, 2014).

The conclusion from these comparisons is that while the top US students in the top US schools at all levels are likely to be globally competitive, the rest of the students in the US distribution are not being served as well by the rest of the US school systems as the students in the other 15 to 20 top ranked countries.

Are there reasons that an industry that is so crucial to the economy and the entire society has remained stuck in a performance rut or is even falling behind each year. Of course, one could argue that much of the technology innovation that has benefited the productivity of the manufacturing and agriculture industries did not translate well to the education industry. Another reason could be that there is no effective market model for education. Most education at the elementary and secondary levels are government organizations paid for with tax receipts. A small number of private schools in each school district have been able to survive despite a cost disadvantage (private school tuitions are usually on top of the public school tax paid by the private school student's parents) as well as regulatory and accreditation hurdles.

There are emerging trends that indicate technology could have an impact on the education industry in the future. For-profit schools that offer online courses and full pathways to high school diplomas have been founded over the last decade and appear to be growing in popularity as a home-school alternative to big public primary and secondary education. Other trends such as "distance learning" (earning a degree from anywhere using online courses) and massive open online courses (MOOC) are being offered by a growing number of colleges and universities as part of their service offerings. While showing some degree of innovation, these new offerings have not yet begun to "move the needle" on productivity in the education industry nor in the quality of education. There has not yet been a Nobel prize winner whose degree is from the University of Phoenix.

Another challenge beyond how to best harness AI technology for the education industry is how to organize education activities so that the social, psychological, and maturity development that usually comes from human interaction can be achieved even if technology can replace the stand-up lecturing function. New terms such as collaborative learning and experiential learning have been coined to give identities to new learning processes that involve students and mentors in learning activities that add to or complement rote training procedures.

It is clear that as automation increases in a greater of number of industries, the need for retraining and continuing education to prepare for the new and displaced members of the workforce will become ever more important. The need for disruptive innovation to meet the challenges of the future in a crucial industry just increases in urgency as each year passes.

Healthcare - Improvements Urgently Needed

The healthcare industry is beset with a series of conundrums. Spending on healthcare by any measure is growing (Exhibit 11.6) at a rate unsustainable for the economy and for individuals as they age. Providers are incentivized to increase the volume of services rather than increase the quality of healthcare. Payers are incentivized to decrease the price of services rather than to increase quality. Patients are stuck in a bewildering web of provider specialists, growing pharmacological dependency, fuzzy and conflicting sparse data bases, long service times, and growing cost burdens.

In addition, the need for healthcare is age dependent (greater need before age 18 and after age 65). The quality of healthcare has a high variation. The skill and training of the care providing professionals have high variation. The organizational effectiveness of care providing facilities have high variation. While the proven success of six-sigma and best practices quality procedures are well-known and implemented in the manufacturing industries and other service industries, they have not been generally adopted by the medical industry for a variety of management and legal reasons.

EXHIBIT 11.6 HEALTHCARE INDUSTRY SPENDING

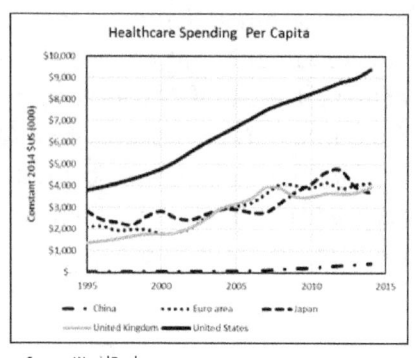 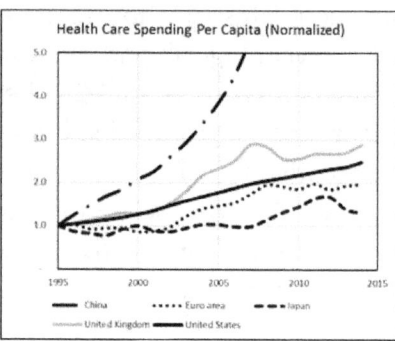

Source: World Bank

Other interesting facts include that half of the spending on health care is spent on 5% of the population, healthcare spending per capita in the US is twice the amount spent per capita in Europe and Japan, that the number of hospital beds per 1000 people in the US has declined by 80% since 1960, and that the number of doctors per 1000 people in the US has increased by over 100% since 1960. Not much productivity to see here.

There is little consensus on what measures of productivity should be used. The simplest is the measurement of the cost to deliver a specific service code. The most complex is to measure the quality of outcome (independent of service) to cost. Since the cost of healthcare in general is such a significant economic problem, productivity measures addressing the cost to deliver a service code will likely be the focus of innovation opportunities for the foreseeable future. However, innovation in alternative diagnostic and treatment technologies as well as continuing fundamental research into the life sciences keep alive the promise that amazing medical breakthroughs are just around the corner.

The good news is that survival rates for many illnesses has seen much improvement over the last few decades. There has been much new technology developed for the healthcare industry that has improved the ability to diagnose, treat, monitor, and prevent many illnesses. The bad news is that new technology often increases the cost of healthcare while new illnesses appear that defy (or are sometimes caused by) the successful treatments invented for previously known illnesses. When magnetic resonance imaging (MRI) machines were invented 30 years ago, it was anticipated that MRI technology would provide new insight into the physiological processes in the human body and would lead to a wide range of new diagnostic modalities. It was once predicted in the mid-1980's claiming that a human ring sized MRI device would be in everyone's medicine cabinet. It was hypothesized that the MRI ring could be used to monitor a person's general health, provide early warning indicators of disease, and eliminate the need for any new x-ray machines. Instead, the MRI has evolved into an expense augmentation to x-ray imaging and CAT scans. And the number of x-ray machines and CAT systems have continued to grow despite the increase in the number of MRI machines. Much like the forecast that the word processor would eliminate the need for copying machines, another hyperbolic technical conjecture bit the dust.

What is needed is innovation that produces better outcomes at lower cost. All of the digital technologies and life sciences are surging in many different directions seeking innovations that can contribute to solving this primary need.

For example, the US healthcare process can be broken down into the following types of services: Physicians and Clinics, Hospital Stays, Home Care, Retirement/Assisted Living (Exhibit 11.7).

In the US, the number of patients per primary care physician (PCP) averages out to about 2,000 assuming the physician spends about 15 minutes per patient visit, patients see a physician about 4 times a year. On an average visit, the patient's basic physiological measurements are taken by a nurse, the doctor reviews the medical history in the patient's file, reviews any recent lab tests, talks with the patient about their feelings about their current health status, makes a diagnosis (if it is not a well-visit), prescribes a course of treatment or a course of drugs, notes his observations, diagnosis, and prescriptions for the patient's history, for prescriptions to be sent, and for invoicing the patient or patient's insurance company. Then the patient leaves the physician's office, goes to buy any prescribed drugs, makes appointments for any prescribed follow-on treatments or tests, and does not communicate with the PCP again until the next visit is required.

EXHIBIT 11.7 HEALTHCARE SPEND BY SERVICE TYPE

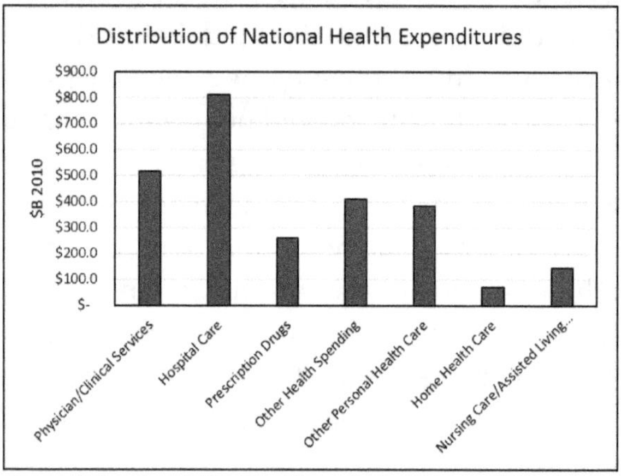

[Bar chart: Distribution of National Health Expenditures, $B 2010, showing bars for Physician/Clinical Services (~$500), Hospital Care (~$800), Prescription Drugs (~$250), Other Health Spending (~$400), Other Personal Health Care (~$400), Home Health Care (~$75), Nursing Care/Assisted Living (~$150)]

Source: US Department of Health and Human Services, Kaiser Family Foundation

Analyzing this generic model of a PCP appointment, several opportunities for improvement using AI and other digital technologies can be envisioned. First is the patient medical history. Most people do not have a complete unified medical history that can be provided to any healthcare provider that needs to see it. If a person has been to more than one doctor in more than one healthcare system, has had lab tests done by more than one lab, has purchased prescribed and non-prescribed drugs from more than one pharmacy, or has failed to take all prescribed medications, then the person's medical history is in multiple pieces on multiple information systems and may not have been recorded at all. It is up to the patient to honestly and accurately tell their doctor at each visit if there is any new information to be added to the record that the doctor being visited happens to have. Unfortunately there is a wide variation in the honesty and accuracy for the quality of the medical history information being told to the doctor by the patient.

A second opportunity where there need to be innovation is the diagnostic process. The quality of the medical diagnosis made by each physician is a function of the physician's training, memory, awareness of new information, and alertness during the examination. Ideally, one can envision an AI system (let's call it MedBrain) that has access to all medical diagnostic information known to the medical profession. MedBrain could then take all the data about patient history, patient symptoms, lab test results, PCP observations, nurse observations, and digital images of the patient, apply data analytics algorithms tests with the result being a list of possible diagnoses rank ordered by calculated probability and a list of treatments rank ordered by quality of outcome for each. Depending on the range of probabilities on the list of possible diagnoses, the final decision would still be up to the PCP and/or the patient. If MedBrain is used by every healthcare provider and facility and all the MedBrains in the world are networked, then the quality of each diagnosis for each patient would be uniformly maximized. And with each patient encounter, the network of MedBrains would continue to learn and add to the global body of medical knowledge.

Our network of MedBrains should increase the productivity of primary care physicians by changing their duties from repetitive service tasks to patient counselors, big data analysts, knowledge managers, process managers, and innovators in of new diagnostic and treatment concepts. The number of patients per PCP could increase dramatically while the quality and cost of this part of the health care expenditure could decrease. It is not hard to image a kiosk or mobile app that could be the most frequent interface between a patient and the entire universe of medical knowledge. People are already becoming familiar with such visions as they search the internet for medical information from one of the many websites already online in advance of or immediately after a visit to their physician.

A third opportunity for cost reduction and productivity improvement would be the use of robotic devices to augment or replace the labor intensive activities now performed by nurses, orderlies, assisted living caregivers, and other healthcare workers. Activities that are amenable to robotic replacement in the near term include medicine and meal distribution in caregiving institutions, monitoring of patient vital signs and behavior. Surgeon assisting surgical robotics have already been successfully introduced to the market place.

The barriers to accelerating innovation in the healthcare industry come from professional, legal, industry structure, and regulatory sources. The evolution of the healthcare industry from the academic training of physicians and nurses, the experiential training enforced by the industry, the method of payments by insurers and the government, and by the legal issues of privacy and malpractice litigation all represent significant structural barriers to change. The best chance is that as new technology offers the promise of overwhelming benefits and as people become more comfortable with AI interactions from their mobile devices, then the barriers can be overcome by commercialized innovation.

Government - Always in Need of Improvement

If any part of the US economy defies productivity measurement, it is the government sector at all levels (local, state, and federal). While identifying the inputs in terms of labor hours or expenditures for each government agency is available to the public, describing the quantity and quality of the output, just as in any organization that delivers services, is not easily done, requires much interpretation, and is not easily measured.

The share of the total economy claimed by all government sectors has grown consistently over the last 50 years to the point where total government expenditures represent 38% of the GDP while total government employment of 22 million people represents 15% of the US workforce (Exhibit 11.8).

Exhibit 11.8 US Federal Government Staffing

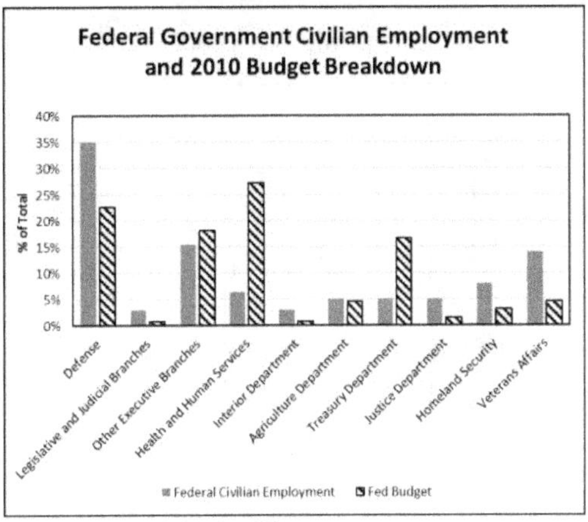

Federal Government Civilian Employment and 2010 Budget Breakdown

[Bar chart showing % of Total for Federal Civilian Employment and Fed Budget across categories: Defense, Legislative and Judicial Branches, Other Executive Branches, Health and Human Services, Interior Department, Agriculture Department, Treasury Department, Justice Department, Homeland Security, Veterans Affairs]

Source: Congressional Budget Office

Not only is the size of the government work force and level of spending the largest in the US economy, there is the persistent belief among many taxpayers that government employees do not work as hard, as long, or just not as efficiently as private sector employees. The data shown from a time-use study (Jason Richwine, 2012) in Exhibit 11.9 indicates that government employees on the average work about one month less per year than private sector workers. Of course, this is just one particular analysis of a set of survey data collected by the US Bureau of Labor Statistics, but it does support the conventional wisdom that there is something about government work that brings out the slothful tendencies of people.

One additional driver of the need for productivity improvement in the government sector is the heavy burden that future pension commitments pose for so many cities and states. Over the last 15 years, over 50 municipalities have declared bankruptcy (e.g. Detroit, MI; Stockton, CA; Harrisburg, PA) with several other cities forecasting insolvency (e.g. Dallas, TX; Houston, TX; Providence, RI) if adjustments are not made to their spending patterns and pension commitments. Puerto Rico reached the point of insolvency in 2016.

EXHIBIT 11.9 EMPLOYEE WORK TIME COMPARISON

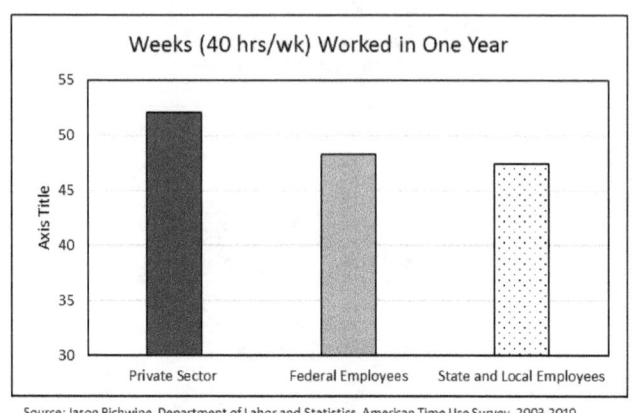

Source: Jason Richwine, Department of Labor and Statistics, American Time Use Survey, 2003-2010

As with most industries, the opportunities and need for productivity improvement in the government sector should be evaluated on the basis of business process flow. Evaluating process flow will help (1) identify how to reduce the cost base, (2) how to redesign workflow, (3) how to apply productivity improving technology, and (4) how to modernize services.

As part of the analysis of each business process flow analysis, it is helpful to answer several key questions.

Question 1: What is the purpose of the business process and how does it rank in value or priority compared to other business processes in the same or other government organizations?

Comments: Quite often, answering this question will result in major savings in cost and improvements in organizational performance. If the underlying purpose for a given business process has become obsolete, ranks low in priority, or duplicates some other process, then it may be possible that the entire process and all the costs associated with it can be eliminated.

For example, if there is a interoffice mail process (that collects, sorts, and delivers envelops and packages) and an external mail process (that sends, receives via USPS, FEDEX, UPS, etc.), it might be possible to eliminate the interoffice mail process entirely by using e-mail for memos and letters between internal users and using the external mail process for anything that has to be physical.

Question 2: Are employees in the process responsible for and rewarded for the quality of outcomes from that business process?

Comments: If there is no system for measuring, identifying, incentivizing, and rewarding the quality of work performed by an employee and having it connected to the quality of the outcome of a business process, then the variance of outcomes is virtually uncontrollable.

Question 3: What can actually be measured as a means of determining productivity in within the business process?

Comments: This is the most difficult and most important question to answer. If the key parameters in a process cannot be measured, it is uncontrollable and subject to the random variance in the behavior and self-motivation of the people in the process. At the same time, defining measurable parameters can be difficult. Easily measurable parameters might be the waiting time of the customers at a department of motor vehicles or patients at a Veterans Administration hospital. Not so easily measurable parameters might be the efficiency with permits are reviewed for approval or how school boards operate. Nevertheless, it can be done. And the process of defining the details of a business process usually reveals which parameters can be measured and which are important.

Question 4: How can plans and strategies to improve the productivity of the government business process translate into an implementation plan that actually delivers on desired outcomes?

Comments: Even after a lot of work has been done by a business process analysis team to identify, prioritize, and reengineer all the activities in a process, creating the plan to implement changes can be organizationally traumatic. Change usually means eliminating, redefining, and/or automating specific tasks or jobs; redesigning job content, forms, information, and work flow; and retraining, firing, and/or hiring people all while transitioning from the old business process to the new. Planning for and managing the change in work flow and people's jobs is an essential part of any productivity improvement program.

Making productivity improvements in service organizations is a challenge in general, but for government organizations, it is especially challenging because of political factors internal and external to the organization. Either a competent management team or a major forcing function (a natural disaster, a budget dilemma, a landslide election, etc.) is needed to overcome the barriers created by these challenges.

Energy - Improvements are Happening

The need for more productivity in the energy sector is driven by any economy that is growing and producing a higher standard of living. Since the industrial revolution, a growing standard of living (more food, more transportation, more retail goods, more healthcare, more entertainment, more social services) has always been accompanied by a growing need for energy to produce and deliver the growing list of goods and services. If the developing world achieves the same standard of living as the US and Europe, the amount of energy that would be needed is twice the amount being produced commercially today. As the direct and implied cost of energy has increased, there are now incentives to reduce the cost of acquiring energy and to reduce the negative environmental impact that those processes create.

The energy industry is one portion of the economy where there has been a consistent improvement in productivity over the last several decades (see Exhibit 11.10). Since 1990, the growth rate of the US economy has been higher than the growth rate in the amount of electrical energy used. In the fact, for the last 10 years, the growth rate in the consumption of electricity has been flat with not growth in demand. This flattening of demand is the result of both an increase in productivity in the use of electricity and an increase in the structure of the economy.

Industries That Need Major Improvements

EXHIBIT 11.10 US ENERGY PRODUCTIVITY TRENDS

Energy Productivity Improvement Trend

Period	Electricity Used	GDP
1950-1990	5.9%	3.8%
1990-2007	1.9%	3.0%
2007-2016	-0.3%	1.2%

Sources: US Energy Information Administration, Dept. of Commerce

One source of the productivity improvement in the use of electric energy come from greater efficiency in electronics, electrical, machinery, and in lighting (See Exhibit 11.11). A second source is in the structure of the economy as it has changed from a manufacturing dominant economy to a service economy.

EXHIBIT 11.11 ENERGY PRODUCTIVITY DRIVERS

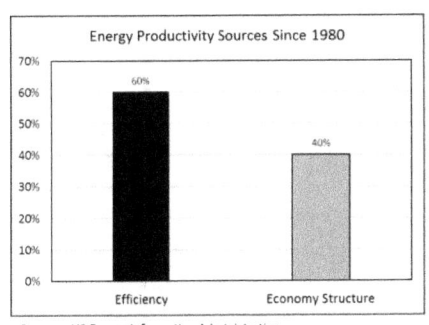
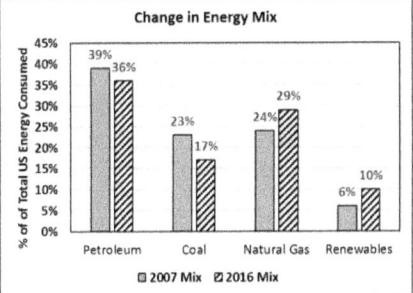

Sources: US Energy Information Administration

The changes in the mix of energy sources as well as in the types of energy uses are also contributing to changes in the structure of the economy. The mix of energy sources for the US in 2015 is indicated in Exhibit 11.12. While it is clear that the transportation sector gets most of it energy from petroleum products such as gasoline and diesel fuel, that ratio could be changing significantly in the next few decades if the anticipated growth in electric car transportation materializes. While the conventional wisdom is that if all cars and trucks were electric, there would be less carbon released to the atmosphere. However, If all gas powered cars and trucks in the US converted to electric power, the amount of power that would have to be delivered by the US power grid would have to almost double. Put aside the fact that enormous amounts of new investment would be needed to build all that new capacity in the current electric power grid, then there is the question of how to dismantle or repurpose the existing petroleum infrastructure. Looking at the new electric power generation required, then the questions would be about which type of energy source would be used by that new generating capacity. If renewables continue to be cost competitive, there will be a significant restructuring in the electric power generation and distribution industry.   The interaction between technology, investment capacity, and return on investment will likely determine how much and how fast growth of this transformation will take.

Exhibit 11.12 US Energy By Source and Sector

Quad = One Quadrillion British Thermal Units (10^{15} BTU's)	Sector (Quads by Source)				
	Trans-portation	Industrial	Residential & Commercial	Electric Power	Total Quads
Petroleum	25	8	1	0	35
Natural Gas	1	9	8	13	28
Coal	0	1	0	11	16
Renewables	1	2	1	5	10
Nuclear	0	0	0	8	8
Total Quads	28	21	10	38	97

Source: US Energy Information Administration

Electric power today is primarily generated in large centralized power plants using either coal, natural gas, utility scale solar fields, or utility scale wind farms (both on land and off-shore). The electric power is then distributed through a grid of thick wires strung from metal towers and wood or cement poles. About 50% of the power generated at the centralized power plants is lost during the generation and transmission through thousands of miles of wire. The last mile of the power grid delivers energy that is used for lighting (15% of total), heating and cooling (35% of total), machinery-appliances (35% of total), and then 15% for electronics and other devices.

One significant trend that is beginning to change and disrupt the utility industry is the growing number of decentralized generation (DG) sites for power. This trend is due to the growing number of installations of sustainable and ecologically beneficial energy sources such as roof-top solar or wind turbines. There are several benefits to an installation base of renewable decentralized electric power generation. The most touted is the clean nature of the energy source (solar and wind). Also important is the elimination of the wasted energy lost during generation at the central fossil fuel power stations and through thousands of miles of transmission wire. A third benefit is that as the power grid becomes decentralized and more networked, there is less chance that a single point of failure in the grid due to a natural or manmade disaster will become a grid-wide shut down. A fourth benefit is that one large natural gas fueled power plant could be replaced by a computer controlled network of a few thousand decentralized power generators (solar panels and wind turbines) that are sitting on the roofs of homes, buildings, parking garages, or standing in the fields of farms and medians of highways.

There are two barriers to a rapid introduction of microgrid technology in the US. The first is technical. A fully capable microgrid needs cost effective and compact energy storage to smooth over the periods when power is not being generated because the sun is not up or the wind has died down. There is a great need for innovation in the battery space specifically or energy storage by other means.

The second barrier in the US is the resistance to change in the power industry by the private and public utilities. Utilities have traditionally had legislated local monopoly positions to ensure that there was sufficient return on the investment made in infrastructure by public or private investors. Now that technology is making a monopolistic position less necessary, there will need to be innovation in public policy and legislated laws that have protected utilities in the past.

Internationally, decentralized power generation (or microgrids as some call such installations) is can be a leapfrog technology. Solar and/or wind powered microgrids can be deployed relatively quickly in developing nations or localities that do not already have long line power transmission infrastructures thus leapfrogging the need to build thousands of miles of drooping wire transmission plant.

An example of conventional thinking about power generation in developing countries is the Grand Ethiopian Renaissance Dam being completed (after a decade of planning and construction) on the Blue Nile River which with another dam recently completed will quadruple the electric power generating capacity of the entire country (Kumagai, 2016). Unfortunately, there is no grid system in place to carry this power from the dam and deliver it to citizens or to export to neighboring countries. It will take another decade or two for the water to fill behind the dam and many billions of additional dollars to build the transmission infrastructure. New power could have been delivered to many communities within just a few months of a start if a solar based microgrid system approach had been pursued.

The trend toward to decentralized power generation will depend on and will create a need for more technology to control the "network" of decentralized power generators as it interacts with and synchronizes with the centralized "grid" of large power plants. As the cost of renewable energy sources continues to decline and as innovations in home automation, smart buildings, smart vehicles, etc., continue to emerge, the demand for more decentralized power generation will continue to grow.

There is a variety of digital technologies that will contribute to the continuing improvement in productivity in the use of energy. These include:

- Artificial Intelligence - Improving the productivity of how energy is used in every sector of the economy
- Smart homes/offices – Improving efficiency of home and office appliances, lighting, heating and cooling

- Grid optimization algorithms - Managing variable renewables with base load and peaker plants
- Solar optimization - Maximum power point tracking for photovoltaic panels
- Storage - More efficient and denser energy storage (more energy in less volume). Includes better batteries, fuel cells for storage, and super capacitors
- Smart materials - Materials which respond to, store, or create energy
- LED lighting - 10X in efficiency and 20X product life
- Lighter, stronger, multi-functional
- Autonomous cars – Could reduce the need for on-road vehicles by 20% to 50%

Innovations in renewable energy will continue to drive down the cost of wind and solar power to make it more cost competitive with fossil fuel based energy. At the same time, innovations in the fossil fuel industry that are driving down fuel cost and reducing its environmental impact.

All these technologies together point to much more productivity to be gained from the production and use of energy in economies of all types.

Water - Many Drops to Drink

The need for more productivity in the production and use of water is driven by the same economic forces driving the use of energy. Every economy that is growing and producing a higher standard of living is also producing a growing need for and use of water. If the developing world achieves the same standard of living as the US and Europe, the amount of water that would be needed is several times that being used today.

Water is a commodity that is economically treated very differently than energy. Because water is a resource that literally falls out of the sky, it has been viewed historically by most cultures as something that is God-given and therefore should be made available to citizen and business alike in the largest volume possible. Whereas, fossil fuel has to be pulled out of the ground at great expense, water is the most plentiful resource on earth. This has led to an economic system that has historically undervalued water and overlooked it as a precious commodity. It is expected that there will be a water spigot or hydrant on every building or at least on every street corner.

The global demand for potable water has become critical because of significantly larger populations, more water intensive industries, and more consumption per capita in high standard of living economies. New productivity improvement technologies that provide for greater conservation, more efficiency, and more recycling in the use of water will be essential for global standards of living to improve.

Water today is primarily produced in large centralized systems that either store water behind dams in lakes, in large ponds, extract it from wells or rivers, and sanitize it for distribution through a grid of pipes, valves, and pumps to individual users. About 40% of water sent through canals, aqueducts, and underground pipe distribution systems is lost to leakage and evaporation. The breakdown of the uses of fresh water in the US in Exhibit 11.13. A couple of observations can be made from this exhibit.

EXHIBIT 11.13 WATER USES IN THE US

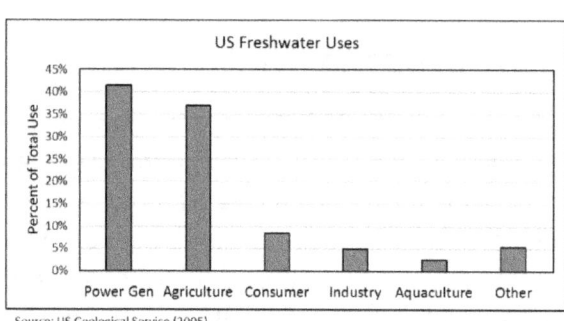

Source: US Geological Service (2005)

First is that since the largest use of fresh water (42%) is in power generation (e.g. water for steam to heat, water for steam to drive electric generating turbines, water through dams to drive generators), renewable power generation from solar and wind has an additional benefit which is that it is less water intensive than fossil fuel power. The chart implies that replacing 20% of fossil fuel electric power generation with renewables doubles the water available for consumer uses.

The second observation is that if the entire fleet of cars and trucks in the US are converted to electric vehicles (as mentioned above) which also means that the capacity of the electric power grid has to almost double, then the amount of fresh water needed might have to increase by up to 50%. This is one of the relationships of energy and water that has led to coinage of the term the water-energy nexus (U.S. Department of Energy, 2014).

A breakdown of the uses of water by consumers indoors is shown in Exhibit 11.14. One far reaching observation can be made from this chart. It is that very little water used by the average person is actually used during cooking, eating, or drinking. Since most of the water flowing through faucets is likely not used for human consumption, the amount of water a human actually digests is a very small percentage of what typically flows through a home or office.

EXHIBIT 11.14 CONSUMER WATER USES (INDOOR)

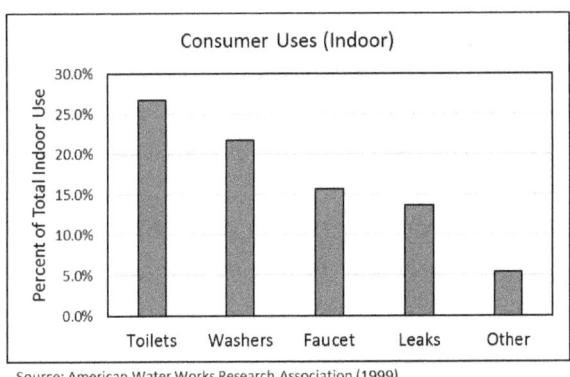

Source: American Water Works Research Association (1999)

Other implications are that there are several things that could be done to free up many times the amount of water that is now used for human consumption. Reducing leaks, making washers more efficient, making toilets more efficient, recycling water used by toilets and washers, or using alternative technologies for toilets and washers are obvious opportunities for improvement.

Digital technologies will provide many of the solutions that can lead to such improvements in the use of water indoors. Making faucets, toilets, and other water valves smart can lead to localized water control technology that stop leaks and reduce water used in toilets and washers. Adding computer controlled purification technology to recycling devices can make the reuse of gray water (toilet and washer water) practical, safe, and reliable.

There is also new digitally controlled technology making it possible to conserve water used in agriculture, landscaping, and industrial applications. All of these improvements help reduce the consumption and cost of water for each economic activity by a few percentage points.

The need for new and significant improvements in recycling technology will likely lead to more decentralization of water systems. Decentralized water recycling will help significantly reduce the water lost in distribution systems and reduce the demand for water volume.

Decentralized water recycling systems will also call for changes in public policies and water regulations. More technology will be needed to ensure that the quality and safety of the recycled water and the effluents from recycling are successfully interfaced into the water grid and sewer grid infrastructures. The theory is that one large centralized water utility can be augmented by, if not replaced by, a computer controlled network of thousands of decentralized water recycling systems that are integrated into homes and buildings.

As developing economies continue to generate larger needs for water due to their growing economies, there will be a growing demand for "leapfrog" technologies in the sourcing and use of water. Decentralized water recycling technology could provide such leapfrog products that will reduce the need to build large new infrastructures and help go directly to the most efficient and effective generation of new water capacity.

Transportation - Disruption Ready

The need for more productivity in transportation technology is driven by the same economic forces driving the use of energy and water. Every economy that is growing and producing a higher standard of living tends to generate a growing need or desire for personal transportation (Exhibit 11.15). If the developing world achieves the same standard of living as the US and Europe, the number of automobiles that would be demanded would be several times the amount being produced commercially today. However, personal transportation vehicles are notoriously unproductive (most passenger cars sit idle 90% of the time) and a major consumer of energy, water, and many other commodities. Mass transit systems, which have been popular in Europe for decades and several of the largest cities in the US, are often seen as a solution for future transportation needs in developing countries as well as in the more rapidly growing urban centers in the US.

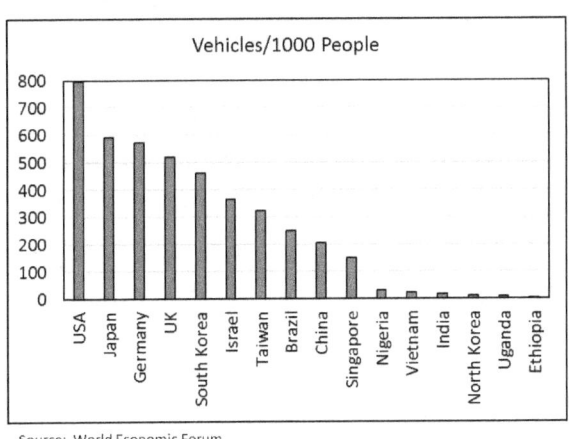

EXHIBIT 11.15 VEHICLES PER CAPITA

Source: World Economic Forum

The reality is that mass transit systems are costly and provide only some of the benefits of an increased standard of living and are not as ecologically gentle as many hope. Load factors (percentage of passengers per seats available) for bus and train systems (Exhibit 11.16) average about the same as personal automobiles (around 20%) even though most automobiles have only one passenger most of the time.

EXHIBIT 11.16 AVERAGE LOAD FACTORS

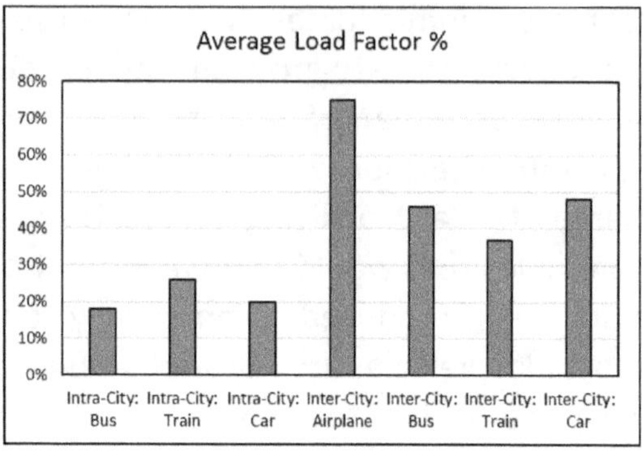

Average Load Factor %

Mode	Load Factor
Intra-City: Bus	~18%
Intra-City: Train	~27%
Intra-City: Car	~20%
Inter-City: Airplane	~73%
Inter-City: Bus	~45%
Inter-City: Train	~38%
Inter-City: Car	~48%

Sources: National Transit Data Base, Bulletin of Atomic Scientists

The cost of mass transit systems are almost always subsidized by taxpayers. The amount of cost coverage provided by the fares that riders pay for public transit systems is usually below 50% (Exhibit 11.17) with a few rare exceptions.

Improving productivity in this sector has been the subject of many government and private funded studies although few practical recommendations are produced. Recommendations from these studies often include imposing restrictions on alternative or competitive transportation modes with the goal of forcing people to fit the physical limitations of the public transit system.

Other types of recommendations include buying bigger vehicles that can be operated by fewer people. Anyone who has ever seen an articulated bus (essentially two buses linked together with one driver) driving around a city mostly empty might wonder what the economics of that proposal looked like in a spreadsheet.

The good news is that there are several technologies that are evolving today that may offer the benefits of higher productivity, higher load factors, lower costs, and lower impact on the environment.

EXHIBIT 11.17 FARES VS OPERATING COSTS

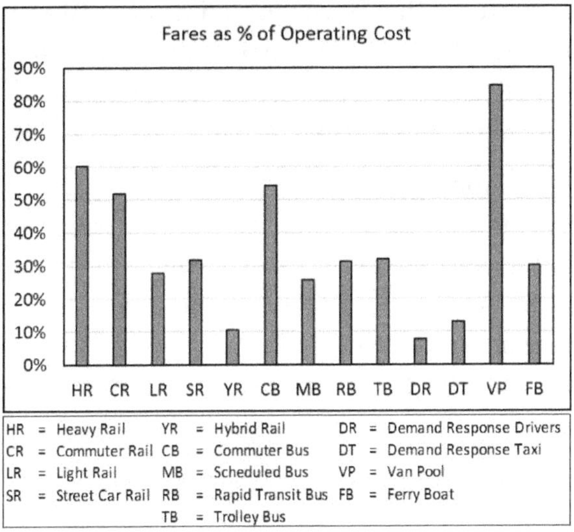

Fares as % of Operating Cost

HR	= Heavy Rail	YR	= Hybrid Rail	DR	= Demand Response Drivers
CR	= Commuter Rail	CB	= Commuter Bus	DT	= Demand Response Taxi
LR	= Light Rail	MB	= Scheduled Bus	VP	= Van Pool
SR	= Street Car Rail	RB	= Rapid Transit Bus	FB	= Ferry Boat
		TB	= Trolley Bus		

Sources: National Transit Data Base

Computer-controlled autonomous vehicles are being developed and tested by several technology companies (Google, Apple, Microsoft), new automotive ventures (Tesla), as well as conventional automotive companies (Ford, General Motors, Daimler Benz, Volkswagen). Autonomous vehicles offer the potential benefits of greater safety, greater efficiency, and less traffic congestion (software that eliminates traffic wave jams by eliminating individual car stops and starts). Greater safety comes from car driving computers that are programmed to look out for each other. Greater efficiency comes from fuel optimization algorithms running in the car driving computers. Traffic jams can be eliminated by software that smooths out traffic waves by minimizing the starts and stops of each car. If all the car control computers on a highway could communicate with each other and agree to travel at 50 mph at a constant separation of 10 feet of separation and to all slow down or speed up simultaneously, then a long line of autonomous cars operates as a virtual autonomous train.

Add to the autonomous vehicle innovations the innovations in software dispatching systems. These mobile app friendly software systems which call for and dispatch for-hire vehicles have been pioneered by several private ventures (e.g. Uber and Lift) around the world. These ventures that have created what is often called the on-demand economy. While they have disrupted the taxi and limousine industry around the world, they are also leading to revolutionary new concepts about personal transportation.

Computer simulations (Exhibit 11.18) indicate that if an entire fleet of cars owned by individuals in a city were replaced with software dispatched autonomous vehicles, the number of vehicles in the fleet and on the street could be reduced by more than 80%, traffic accidents could be reduced by even a larger percentage, and energy used could be reduced accordingly (Fagnant, Kockelman, & Bansal, 2015).

EXHIBIT 11.18 FULL AUTONOMOUS FLEET IMPACT

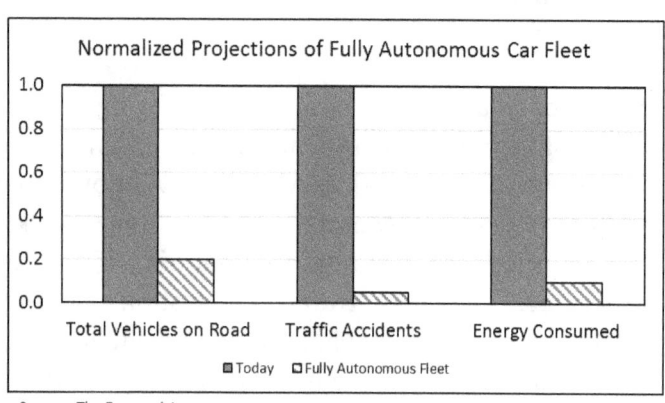

Disruption of and a complete conversion of the automobile industry into an autonomous personal transportation industry has many economic, social, legislative, and government policy obstacles and may not happen at one time or ever. However, the financial and safety benefits that could be produced by the introduction of such technology is intuitively appealing and may be economically compelling.

Barriers have been raised against the forerunner technologies (on-demand ride sharing services like Uber) by labor unions and government officials in many cities because they fear loss of jobs or loss of tax and licensing income. At the same time, there are some cities that are experimenting with on-demand services like Uber to augment or replace their public mass transit systems (Brustein, 2016). In 2016, both Uber and Lyft formed agreements with transit agencies in San Francisco, Atlanta, Philadelphia, Dallas, Cincinnati, and Pittsburgh for services such as shuttling commuters to bus or train stations. Since last year, Uber has scored public transit agreements with San Francisco, Atlanta, Philadelphia, Dallas, Cincinnati, and Pittsburgh. Officials in Washington, DC proposed having Uber respond to some 911 calls for ambulances. The city of Altamonte Springs, Florida created a pilot program using Uber as a personalized on-demand alternative to a city bus system.

This type of innovation, of using a network of decentralized, individually controlled, highly utilized vehicles could revolutionize the transportation industries in the developed countries and provide a potential "leapfrog" opportunity for developing countries that have not invested in massive mass transit infrastructures yet. It could also simplify the investment in roadway systems if they were designed with networked controlled autonomously driven vehicles in mind. For example, physical traffic control signal systems could be replaced with virtual traffic control systems built into the traffic network control and autonomous vehicle systems.

It appears that for the foreseeable future, the evolution of the transportation system will be a blended combination of human controlled and operated, and machine controlled and operated vehicles. Also, there remains considerable technical work to be done to make the AI safe and stable for systems where lives are at stake. Anyone that has ever rebooted their laptop or cell phone knows that the state of the technology available to consumers is somewhere short of infallible. This evolution will likely lead to changes in government licensing laws, car insurance policies, and liability tort systems. The combination of emerging technology and significant productivity and safety benefits could signal a global tipping point.

Other Service Industries

The data in Exhibit 11.19 should dispel any doubt about the need for productivity improvement in the services sector. Since services represents about 70% of US GDP, the level of productivity in the service sector is the biggest driver of productivity in the economy.

EXHIBIT 11.19 SERVICES VS MANUFACTURING PRODUCTIVITY

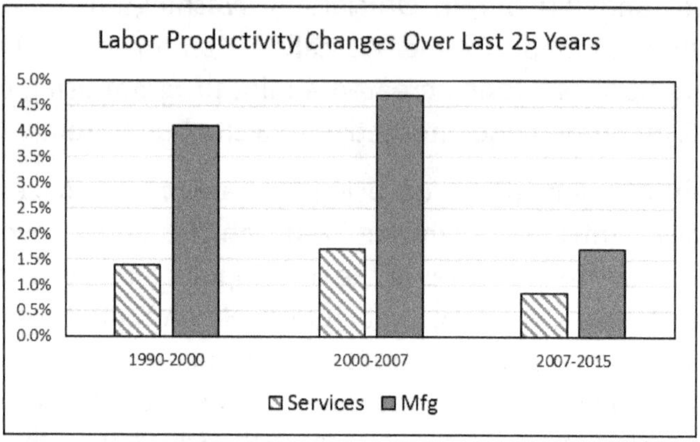

[Bar chart: Labor Productivity Changes Over Last 25 Years, comparing Services and Mfg for periods 1990-2000, 2000-2007, and 2007-2015]

Source: Bureau of Labor Statistics

For the last 25 years, productivity improvement in the service sector has been one half to one third that of manufacturing. Even during the period post the 2008 recession, services productivity improvement has remained one third that of manufacturing.

Breaking down the services sector (Exhibit 11.20) shows which services are struggling the most. Traditional service industries such as department stores, restaurants, and grocery stores have shown only small improvements at best. The service industries that showing the greatest gains in productivity are also currently the fastest growing segments: Electronics stores and E-commerce (online retail, business-to-business supply chains).

It is not only the domestic retail industry which is feeling the effects of disruptive technology. While technology is causing changes to established infrastructure in the US, it is helping leapfrog infrastructure barriers in other countries.

Industries That Need Major Improvements

EXHIBIT 11.20 INDUSTRY SEGMENT PRODUCTIVITY

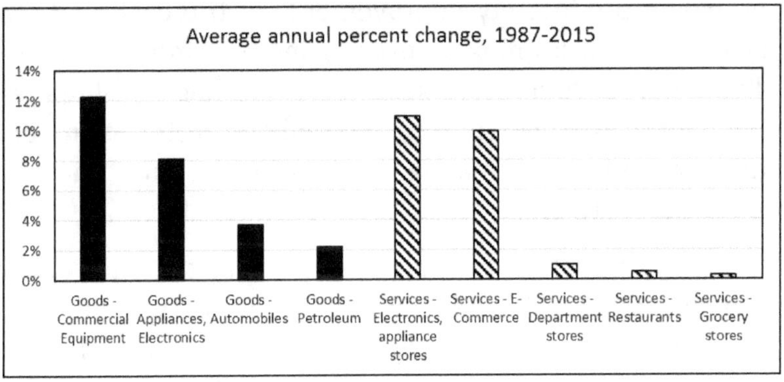

Average annual percent change, 1987-2015

Segment	Approx. %
Goods - Commercial Equipment	~12%
Goods - Appliances, Electronics	~8%
Goods - Automobiles	~4%
Goods - Petroleum	~2%
Services - Electronics, appliance stores	~11%
Services - E-Commerce	~10%
Services - Department stores	~0%
Services - Restaurants	~0%
Services - Grocery stores	~0%

Source: Bureau of Labor Statistics

In the US, decades of expansion of real estate square footage for retail stores is now being reversed as online shopping and order fulfillment from distribution centers is replacing a growing share of sales transactions. This trend has several demand side reasons, one of which is the fact that shoppers find it convenient to shop using search engines on a mobile device and then to have the product delivered to their front door within a relatively short time. Physical retail stores are being transformed into showrooms, entertainment centers, and pickup stations for online ordered goods.

In China, a country whose economy has grown fast and has taken advantage of leapfrog technologies already (such as installing cellular phone technology and foregoing decades of building pole strung wires for landlines), the retail industry is on a path to leapfrog decades of building retail real estate centers (strip malls and shopping centers) and to grow primarily through e-commerce technology and order fulfillment supply chains.

The technology making all this happen in both economies comes from almost every direction: augmented and virtual reality, online applications, content management software, wireless technology, supply chain wide enterprise inventory management systems, automated warehouse and distribution center systems, and networked delivery transportation systems.

Technology is not only supporting trends in shopper preferences, but also providing productivity improvements in the cost of the retail value chain. It is disrupting an existing industry infrastructure with the elimination of retail store sales clerk jobs and retail real estate management jobs while creating new jobs in distribution centers, inventory robotic and management systems, and in delivery transportation network.

It is changing the nature of the industry where the once dominant retail chains like Macy's and Sears are struggling to compete with dominant online retail and supply chain companies such as Amazon, Google, and Apple. The percent of the retail industry that is delivered through e-commerce is small now (about 8%) but growing rapidly (doubling every 4 years) likely unabated especially with leapfrog opportunities in the rapidly developing countries.

In the financial services sector, technology that provided productivity improvement began emerging with the advent of the ATM machine in the 1970's and now continues with online banking, stock transaction, credit card transactions, and other types of electronic money management. New ventures in the financial services sector are creating changes in the established banking industry.

While ATM machines have proliferated, the number of bank teller jobs has not decreased while also many new jobs were created in the supply chain that services the ATM machines. ATM technology decreased the cost of servicing cash transactions for customers dramatically thus making it cheaper for banks to increase the number of local branches. It also changed the work content of tellers from simple cash transactions to more complicated and more value adding financial transactions.

In the restaurant and food services industry, technology has not yet had much of an impact in improving productivity in the order taking, preparation and delivery of meals. However, the success of ATM's, airport kiosks, online reservations, online ordering of food delivery, and other digital interfaces for consumer goods will likely lead to greater use of technology in this industry. As illustrated by the hamburger store example in a later chapter, the speed of the introduction of new productivity improving technology in this industry is a function of investment requirements, customer demand, and return on investment.

Ultimately, the most effective application of technology in the design of any service organization depends on understanding the work and process flow in each of the business processes. Evaluating each activity in each business process and understanding how it meets the performance objectives of the business is essential. This type of analysis helps make it clear where automation should be applied to have the greatest impact on not only productivity improvement but also in customer satisfaction. In some processes, automation may only be applied to specific activities while people perform the remainder of activities. Even in online retail commerce, customers may interact with the software on a website that can carry on a conversation, but there are many people behind the user interface screen managing and updating the software, operating the distribution centers, and driving the delivery trucks that complete the transaction with the customer.

SECTION V

EVOLVING TOGETHER

Chapter 12

MACHINE EVOLUTION

Since the invention of the wheel, each new development in technology and each new application that improves productivity has been a step forward in machine evolution. The first applications of technology that had significant impacts in the lives of people were primarily in replacing or amplifying human muscle in routine activities. This was the golden age of mechanical machinery in the industrial revolution. As the ingenuity of mechanical engineers grew and the efficiency of machinery improved, the replacement and leveraging of human muscle in non-routine activities became more practical.

After the concept of binary digital computers was created and then electronic switches were invented, muscle extending machines became smarter and could be applied to cognitive activities whether routine or not. Combination of electronics and new mechanical materials have allowed technology to scale in size upwards (earth movers, moon rockets, locomotives, etc.) or downwards (pacemakers, thumb drives, cameras) to less mobility (intelligent buildings) to more mobility (smart phones, Rumba robots).

A model of machine evolution is illustrated in Exhibit 12.1. First machine muscle was controlled by humans to perform routine manual tasks. Then machine muscle became more powerful through the use of hydraulic control motors. Along came more efficiency and more power with electric motor controls. In addition to more power, machines became more flexible and could be applied to support non-routine manual tasks or applied to routine cognitive tasks (with the aid of sensors and/or fixturing). With the advent of computer controls for machines, a new era of applying machines to tasks that demanded flexibility and real-time but routine decision making.

Exhibit 12.1 Machine Evolution Model

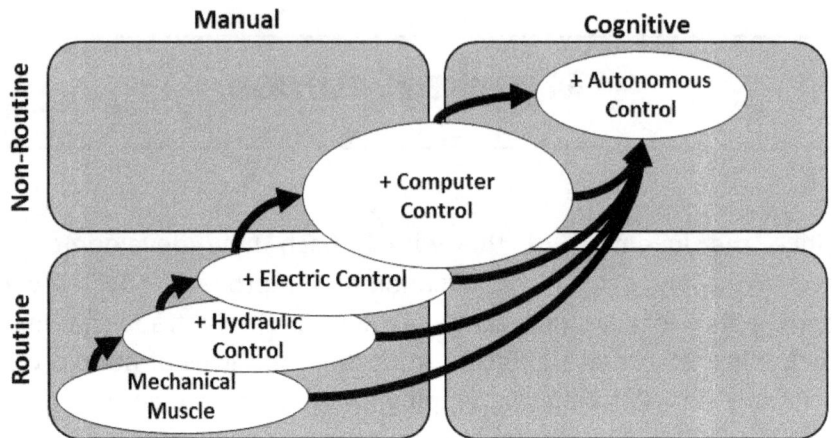

Now emerging are machines that can be controlled by semi-autonomous AI. Theoretically, autonomous AI electronically senses the operational environment (e.g. driverless cars) and operates a machine without human intervention up to a point. That point is usually when the real world provides a challenge to the machine for which it has not been programmed (i.e. it gets confused), when the machine breaks down (and can't fix itself), or when the machine runs out of fuel/energy.

What comes after that? Given the randomness of the cosmos, it could be AI should become more collaborative as well as somewhat autonomous. Collaborative AI provides the people with which is designed to help with advice and feedback about what it knows or senses. Our mythical heroes (whether sourced from Norse gods or comic books) usually have helpful sidekicks. Hunters have faithful and well-trained dogs. Why not have a diagnostic AI system consult with a caregiver (i.e. physician, physician's assistant, nurse, relative) on the management of a patient's situation? A patient today can be monitored on a continuous basis by wearable electronics or a smart home system. Why not have scientists consult with an AI system when planning and performing research projects? Why not have educators consult with an AI system when managing a student's educational programs.

The net effect of technology innovation is that people will remain in control because people design, build, and pay for the digital technology that is needed to improve their quality of life. Machine evolution is guided by people.

Chapter 13

HUMAN EVOLUTION

A model for the evolution of a path for how humans interact with machines is shown in Exhibit 13.1. This uses the same format as that shown in Exhibit 12.1.

EXHIBIT 13.1 EVOLUTION OF HUMAN INTERACTION WITH MACHINES

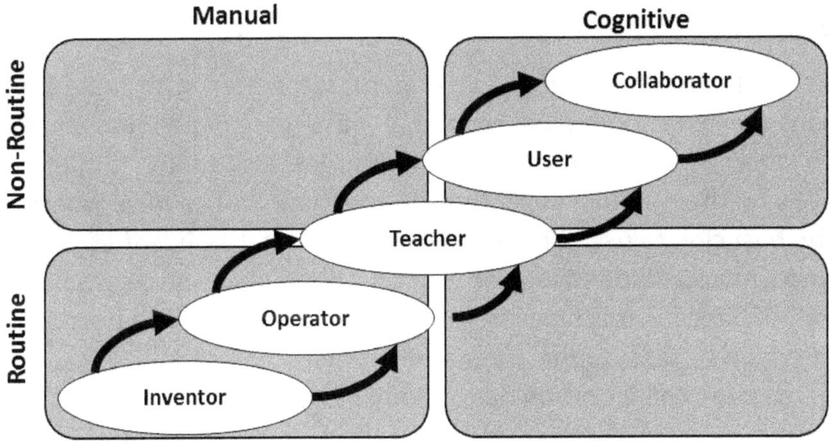

The earliest of the human species invented the most basic of tools. As their brain developed, their knowledge base increased, their experience with the benefits of innovation expanded, and their inventions became more complex and more productive. As machine inventors increased the complexity and power of machines with hydraulic and electric motor controls, humans became skilled operators that required a greater use of their cognitive and motor skills.

As machines became intelligent with the advent of electronic and computer controls, humans became teachers (i.e. programmers) of the machines by having to expand their thinking beyond just the operational problem to be solved by the machine. The teachers/programmers of machines had to put themselves inside of a machine, understand its most basic operations, write instructions for the control computer to direct every single part and activity of the machine. People had to invent the machines brain and central nervous system.

As machines became more intelligent and more autonomous, humans became users rather than operators. People became comfortable "using" an ATM to perform basic banking functions rather than dealing face to face with human bank tellers. People became comfortable with using a website to shop online rather than visiting malls and talking to a sales clerk.

As machines became more intelligent, integrated with multi-media sensors, and connected to global data bases, people began to collaborate with machines rather than just use them. Golfers use apps on their mobile devices or on their golf cart to help determine the best club to use. People talk to their digital assistant at home to find a musical tune to play or turn on an appliance. Surgeons consult with diagnostic machines in real time to determine the best course of action to take during surgery. That first small step into the world of machine collaboration has already been taken.

Is this step toward human-machine collaboration a step forward in human evolution or just another example of tool making? The answer is probably both.

The Flynn effect discussed in Chapter 8 implies that there may be changes taking place in human cognitive capabilities in a relative short amount of time. The fact that research has found that the average IQs of definable populations and nations have been increasing consistently since the 1930's indicates that something physiological, environmental, or observational is happening. Investigators have been searching for over 20 years looking for IQ test biases, contributing factors, demographic dependencies, and genetic factors.

Hypotheses about what factors may be causing the measurable improvement in IQ include improved nutrition, better health care, more years of education, more mental stimulation of children at early ages, and more familiarity with test taking. Although there has been no clear definitive explanation for the upward trend, ongoing testing over the years indicates generally that (1) most of the IQ improvement occurs in the lower half of the bell curve of IQ, (2) populations from less developed countries (lower standard of living) tend to have the lower average IQs but tend to improve the most, and (3) IQ improvement for a population from a developed country (high standard of living) tends to diminish and even stall after many decades of upward change. It is as if there is a limit to how much the standard deviation of the IQ bell curve can be reduced or a limit to how much the "IQ inequality" range can be reduced.

Then there is the relatively new field of study called epigenetics. Epigenetics is the study of changes to the structure of DNA molecules that alter the expression of genes. These changes to the structure are called epigenetic marks, and they differ from mutations in that the actual sequence of a gene remains unchanged. These marks occur naturally and regularly, but can also be influenced by age, lifestyle and environment. One interpretation of epigenetics is that is a physiological mechanism that enables the genome of an individual to sense an environmental change, respond to it, and transmit the response to descendants. If this is indeed the mechanism at hand, it would be evolution at hyper-speed.

Nevertheless, as of 2016, there has been (1) no discovery of an "IQ gene," (2) no clear indication of whether epigenetic phenomena can explain the Flynn effect, and (3) only an initial understanding of how lifestyle or environment might influence human cognitive capacity over only a few generations. However, the story that history tells (Exhibit 13.2) is that as people make machines smarter and stronger, machines help people become more intelligent and more innovative.

EXHIBIT 13.2 SYMBIOTIC MACHINE AND HUMAN EVOLUTION?

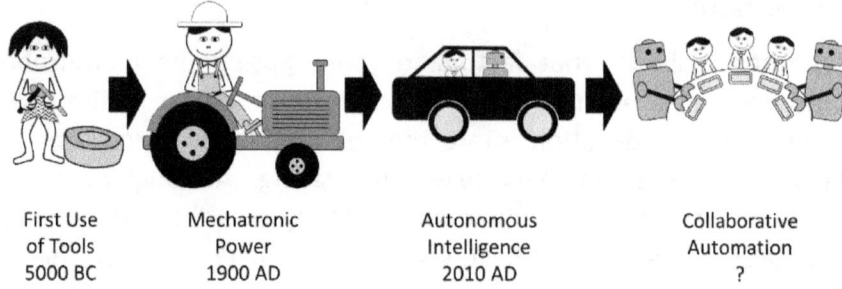

First Use of Tools	Mechatronic Power	Autonomous Intelligence	Collaborative Automation
5000 BC	1900 AD	2010 AD	?

Over the last century as people have innovated in the development and application of intelligent machines, people have had to adapt to greater demands on intelligence, to adapt everyday life to share with machines, and to deal with new social and physical problems even as old ones were eliminated. This evolution of human behavior and ingenuity has been a direct result of and in parallel with the evolution of the technology that it has been developing over the same time. While calling this trend a simultaneous evolution might be a stretch, the implication is rather clear. Humanity's future is intertwined with the future of the machines that it is creating.

SECTION VI

IMPORTANT QUESTIONS AND ANSWERS

Chapter 14

HOW FAR OFF IS THE HORIZON?

Technology Adoption Rates

The rates at which new technologies are adopted by the general economy and become part of everyday life depend on many factors. Capital must be available to invest in the new technology. There must be a need for the technology to create a demand for it. The price of the products made possible by the new technology will determine what portion of a population can afford the new products. The time frame for full penetration can range from decades to centuries. Full penetration may not happen because a technology can be made obsolete by a newer technology.

The printing machine that made it practical to produce reading material was the Gutenberg press which was invented around 1440. It took a century or two before enough printing presses had been built to make books affordable for the small portion of the population in Europe that could read. Affordable books made it affordable and purposeful for people to learn how to read which lead to a greater demand for books. The penetration rate was slow but consistently increasing as these two productivity trends, a decreasing cost of books and increasing literacy rates, reinforced each other.

The US data in Exhibit 14.1 from the more recent history of electrical and electronic products tells several stories. The first is that technologies that are fundamental and require a new infrastructure may require decades.

EXHIBIT 14.1 TECHNOLOGY ADOPTION RATES

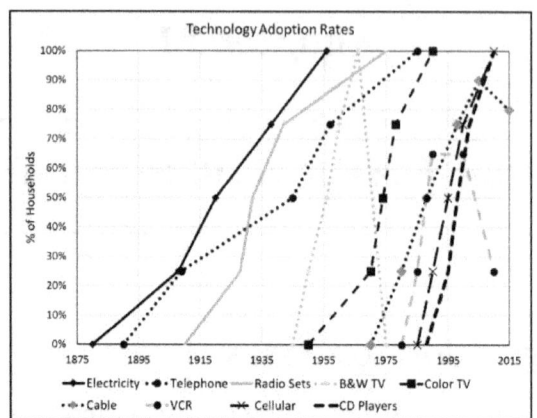

It took over 70 years of intensive capital investment and a national strategy to electrify the entire continental US. This required the construction of power generation plants and hydroelectric dams as well as a national power grid to distribute the energy produced.

Even by building along with and in many ways on top the growing electrical grid, the landline telephone took over 100 years to reach the entire population. Over this time, there were several generations of telephone switching technology that were invented, implemented, obsoleted, and then discarded before 100% coverage was accomplished.

Radio technology did not need as much infrastructure and was able to reach 100% coverage in just over 50 years.

Cell phone technology benefitted from the infrastructure of the telephone and radio industries. Full coverage of the population happened in less than 30 years.

Black and white television was able to penetrate the mass market in about 20 years. It also benefitted from the technology and infrastructure of the radio industry. Black and white television was obsoleted by color technology, but full penetration had to wait for almost 40 years as the broadcasting and manufacturing industries made color cheaper and more appealing to the average consumer and for the installed base of black and white televisions to be replaced.

Consumer electronics products such as VCRs (video cassette recorders) and CD (compact disc) players took less than two decades to achieve significant penetration but were rather quickly made virtually obsolete by innovations in digital streaming technologies.

The lesson learned from these examples is that it is difficult to foresee whether a technology trend will have a significant impact on the economy, how long it might take, and how long it might last. Jared Cohen, president of the Google owned company called Jigsaw, a leader in understanding how digital technology has changed current society, and a frequent speaker at public gatherings about the future of digital technology said that "anyone who thinks they can predict what will be happening in technology more than 5 years into the future is writing science fiction" (Cohen, 2017).

Even though the objective for this book is not to be a work of science fiction, there are a few observations about how long it might take for new digital technology innovations to enter the economy. Innovations usually take years to work its way into the lives of a majority of a population. The question is usually how many decades. If an innovation requires infrastructure changes, then the penetration rate may take several decades. If the innovation is a new mobile app, full penetration rate could take only a few weeks.

Converting the entire personal transportation system over to electric-powered autonomously-controlled cars will likely require infrastructure changes (electric grid expansion, highway modifications, manufacturing industry disruptions, car ownership and financing changes, etc.). On the other hand, there could be an innovation waiting in the wings that could shorten this adoption rate. A car-roof-top solar powered vehicle with enough sensors, software, and computing power to operate safely in a city with the usual number of insane human drivers might be one answer.

Creating new automation and artificial intelligence innovations for the medical and health care industries could require infrastructure changes as well as a turnover of a generation or two of medical practitioners. The power of computer aided design (CAD) systems was brought to market in the early 1980's. It was quickly adopted by the engineering and drafting professions as a way to digitize the creation of 2D blueprints. The 3D power of CAD was vastly underutilized for another 20 years until a new generation of engineers and drafters were educated to think and design in 3D and then entered the marketplace. The same may be necessary for professional services industries like education and healthcare.

It is not necessary to resort to sci-fi expediency to make the point that the challenges of introducing new digital technologies are created by both financial and human factors. A hypothetical case example might help highlight typical challenges.

Thought Experiment: Hamburger Automation

Possibly the one category of commercial establishment that has been visited by the largest percentage of the global population is the hamburger joint (using the vernacular of a child of the 1960's and/or hipsters of the 2010's). Essentially an American invention of preparing comfort food in a short amount of time (i.e. "Fast Food"), the hamburger store first became an American success story in the mid-1900's (e.g. Crystals, White Castles, McDonalds, Burger King). It was a success story not only because of the enjoyment it gave its customers, but also because of its organizational efficiency. These first fast food franchise enterprises became examples of how the quality control of perishable food across a global supply chain could be mastered, standardized, and optimized in terms of cost and taste.

McDonald's has become a global business success in essentially all cultures and geographies. The first McDonald's store was opened on the Champs-Élysée in the 1970's. Then beginning at the start of the new millennium, McDonald's (and fast food companies in general) became a target of the eternally vigilant health conscious. Never mind that the company's success in cost and quality control resulted in a supply of high quality food with high nutrition that met the taste demands of consumers in many countries at relatively low prices.

The topic of whether the epidemic of obesity in the US was due to the existence of fast food chains or due to the lack of dietary self-control of an increasing affluent and decreasingly physically active society is best left to a different forum of discussion. Short of hamburgers being declared unconstitutional, analyzing the economics of how automation could affect the evolution of a fast food franchise make for a useful "hypothetical" case analysis to illustrate the practical aspects of deploying automation.

So for the purpose of this analysis, let's look at a very large hamburger corporation that happens to have a circus clown as its corporate logo and mascot. It is a publicly traded company, so the key operating numbers are available. It is likely that this company has already done numerous studies on what it would take to automate most if not all of the activities in a typical drive thru store.

The appropriate approach to such a study would be to define the work flow of activities necessary to deliver delicious, high quality food to the customer in a timely and attractive price. The work functions would likely consist of the activities listed in Exhibit 14.2.

EXHIBIT 14.2 HAMBURGER WORKFLOW ACTIVITIES LIST

	Work functions				
1	Drive thru order entry	7	Drink maker	13	Supply room distribution
2	Drive thru order delivery	8	Salad maker	14	Kitchen clean up
3	Walkup order entry	9	Special maker	15	Ops Mgr
4	Walkup order delivery	10	Restaurant clean up	16	Financial Mgr
5	Sandwich maker	11	Grounds clean up	17	Inventory Mgr
6	Fry cook	12	Supply room receiving	18	Maintenance staff

The basic business processes might be as illustrated in Exhibit 14.3.

EXHIBIT 14.3 HAMBURGER BUSINESS PROCESSES

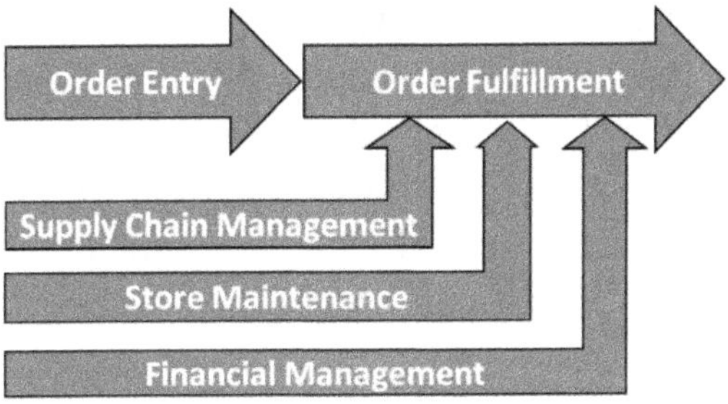

- Order Entry
- Order Fulfillment
- Supply Chain Management
- Store Maintenance
- Financial Management

A summary of a cost model of conventional manual operations versus a fully robotized store is shown in Exhibit 14.4. There are many assumptions included in this model and the weakest of them are highlighted in the following. Nevertheless, the conclusions are revealing.

EXHIBIT 14.4 MANUAL VS AUTOMATION COST MODELS

Employees/Shift (Conventional Store)			Replacement Robot Costs			Hypothetical Hamburger Store Automation Model		
Title	Heads	$/hr	# of Bots	CAPEX $/Bot	Ops Cost $/Bot-hr	Cost per shift per store		
Drive thru order entry	1	$20.00	1	$50,000	$2.25	Manual	$/Shift	$2,522
Drive thru order delivery	1	$15.00	1	$100,000	$4.48	Automated	$/Shift	$984
Walkup order entry	2	$20.00	2	$50,000	$8.97	Savings	$/Shift	$1,538
Walkup order delivery	1	$15.00	1	$100,000	$4.48	Savings	$/yr	$1,119,920
Sandwich maker	1	$15.00	1	$50,000	$2.25	Investment in Robots	$	$1,035,000
Fry cook	1	$15.00	1	$75,000	$3.37	Shutdown/Transition	Months	6
Drink maker	1	$15.00	1	$25,000	$1.14	Lost Ops Margin - Shutdown	$	$552,436
Salad maker	1	$15.00	1	$50,000	$2.25	Total Cost to Automate Store	$	$1,672,356
Special maker	1	$20.00	1	$75,000	$3.37	Payback period	Shifts	1,087
Restaurant clean up	1	$15.00	1	$50,000	$2.25	Payback period	Years	1.5
Grounds clean up	1	$15.00	1	$50,000	$2.25			
Supply room receiving	1	$20.00	1	$100,000	$4.48	Annual Converted Stores Rate	Stores/Yr	1,076
Supply room distribution	1	$20.00	1	$100,000	$4.48	Using Current CapEx Rate		
Kitchen clean up	1	$15.00	1	$100,000	$4.48	Time to Convert all Restaurants	Years	34
Ops Mgr	1	$45.00	0	$ -	$ -	Total Cost to automate	$B	$61
Financial Mgr	1	$30.00	1	$5,000	$0.47			
Inventory Mgr	1	$30.00	1	$5,000	$0.47			
Maintenance staff	1	$40.00	0	$ -	$ -			
			Total	$1,035,000				

The first assumption is about the cost of the robotic equipment that might be used to replace the employee activities for each activity. Making assumptions about each type of robot for each work category totals up to be approximately $1 million. The problem with this assumption is that it is probably too low. First of all, many of the robots that are hypothesized to replace each work function do not exist and have not been invented yet.

The second is that the total cost of all the current generation kitchen equipment that would be installed in a new McDonald's restaurant in 2017 ranges between $0.5M to $1.0M. Current generation kitchen equipment has the benefit of decades of learning curve and volume manufacturing. It is likely that a fully automated kitchen and food service delivery robotic system (which does not exist yet) would be several times the capital cost of a conventional manual kitchen. So count the $1M estimate for hamburger restaurant automation optimistic.

A second assumption is that an existing restaurant could be shut down totally renovated, fully automated, and back up to full revenue operation in 6 months. This means that 50% of the annual variable operating margin (revenue less cost of goods and labor) of the restaurant would be lost. Any savings from the implementation of the automation would have to cover the capital expenditure (CAPEX) on the renovated restaurant and the lost variable operating margin from the shut down and transition period.

The 6 month number might be conservative as well if the recent experience McDonald's has had with the introduction of order entry and payment kiosks stationed at the entrance inside restaurants is any indicator of the reluctance of the customer base to immediately come back pre-shutdown sales volumes. My experience after visiting 5 McDonald's restaurants in 3 cities over a 6 month period in 2016 revealed what appeared to be the first wave of customer facing automation crashing on the rocks of the digital shore. For the first month of operation, there would be a restaurant employee standing next to the kiosk to explain its operation, help a customer make a transaction, or to perform the transaction for the customer. By the third month, the employees behind the normal counter were waving customers on over to their positions. By the sixth month, the kiosks were turned off.

Added to the obvious shortcomings of this model that it might be optimistic in the estimates of cost and availability of the automation needed, is the fact that there was no estimate of what investment might be needed to maintain, upgrade, and improve the system. Leaving this out makes the model even more optimistic. The technology would likely not stand still for 34 years.

The model implies a cost savings of about $1.1 million per store which implies that the payback period is about 1.5 years. There are 36,000 restaurants in the McDonald's global operations. The total CAPEX for the company in 2015 was $1.8 billion. If the company spent all of its annual CAPEX on converting its restaurants to full automation, then just over 1,000 restaurants per year would be converted per year. At this rate it would take 34 years to convert all the restaurants in its operations to full automation and a total of over $60B in capital investment. And this is the optimistic version of the model. The payback period for this analysis might be appealing to the final decision maker, but the total capital commitment required and the time frame for full implementation would likely not be.

This simple example highlights the types of challenges that have to be surmounted before a decision to invest in the deployment and implementation of automation can be made. This is just one simplified model for one simple portion of a complex and dynamic economy. As was mentioned above, forecasting technology farther than 5 years into the future is science fiction.

Chapter 15

DOES THE ECONOMIC SYSTEM MATTER?

The Short Answer: Yes.

It matters in terms of how investment is made in the economy, what institutions exist to support innovation, how much automation is deployed, how fast it gets deployed, where it gets deployed, who benefits from the deployment, and how the disruptions from deployment are managed (Acemoglu & Robinson, 2012).

Since the type, speed, and area of deployment of automation depends on investment, how decisions about investment get made becomes the ultimate driver. Since there are variations of economic systems, there are variations of decision processes on capital investment.

Going back a millennia or so, the prevailing economic systems were feudal in nature and centered around a few villages, cities, or regional kingdoms. Decisions about investments were made according to rank on the official royal social order or rank according to size of holdings on the list of land owners. Trade between the villages and cities was fostered by entrepreneurial traders that invested in horse drawn wagons and carriages.

Then during the middle part of the second millennium, colonization of the African, North and South American, and Asia continents by the naval empires of the European continent created a variety of government dominated, top down controlled network of local market based economies that were tethered to the headquarters of the empire in charge at the time. Most trade outside of the local economies went to the European country that created the empire.

The American revolution in the late 1700's created the first combination of a democratic government and a market based economy. This launched the modern era of experimentation in combinations of national government structures and economic systems.

The American national model is based on a market based economy where investment decisions are made by business entities in response to the forces of supply and demand within the borders of the nation state. The role of government in this model is primarily to ensure that the markets are fair, safe, and a source of revenue for the government. Over its history, the size and role of the government has expanded to include financial protection and stimulus to the markets and to implement some form of wealth transfer from those that have it to those that do not have so much.

Other combinations of government and economic systems were tried in the 20th century. Communist governments that resulted after revolutions in Russia and China implemented economic systems that were centrally planned by the national government headquarters. All prices, production levels, and product/service choices were set by the central governments. Both of these systems failed. The Soviet empire failed after 70 years. The communist Chinese centrally planned economy turned itself into a market based economy after 40 years of existence. Both of these events occurred in later-half of the 1980's. Although the political systems of both former communist empires have moved a bit closer to a democratic system, both are still basically one party systems where opposition movements are discouraged or functionally stymied.

While the domestic economic systems of most developed countries are now market based, it is the growth of a market based system internationally that has emerged the fastest during the last 40 years. First it was the Japanese economy in the 1970's and 80's that focused on exports of low cost and high quality goods to the US and Europe. They were followed by South Korea, Taiwan, and eventually China. It is now China that has grown its economy to a massive scale based on exports to the US, Europe, and even Japan. While Japan's domestic economy has matured to the point of stalling for the last two decades, the domestic Chinese economy has been growing rapidly over the last 10 years. The Chinese economy started from a small base with much room for further growth before it reaches anything like the maturity of the US, Japanese, or European economies.

Although the developed economies are all market based now, there are a range of factors that determine their competitiveness within the global market (Exhibit 15.1). The factors include degree of innovation, availability of venture capital, relative size of exports and imports, regulation burdens imposed by governments, tax rates, government debt, and personal savings.

EXHIBIT 15.1 COMPETITIVENESS RANKING OF TOP ECONOMIES

	Global Competitiveness	Innovation	Venture capital	Exports % GDP	Imports % GDp	Foreign market size	Domestic market size	Regulation Burden	Tax rate	Gov Debt % GDP	National savings % GDP
Switzerland	1	4	13	9	11	15	14	3	6	8	5
Singapore	2	6	3	1	1	9	15	1	4	18	1
United States	3	9	1	21	21	1	1	15	14	19	18
Finland	4	5	4	12	10	20	19	2	12	9	20
Germany	5	3	14	10	9	2	3	12	16	13	10
Japan	6	7	12	20	19	4	2	13	17	21	13
Netherlands	7	15	11	4	3	6	10	7	11	12	6
United Kingdom	8	17	10	15	13	5	4	9	8	15	21
Sweden	9	2	6	11	12	16	12	5	18	6	9
Norway	10	14	2	13	18	18	17	10	13	2	2
Denmark	11	8	20	8	8	17	18	14	3	7	11
China	12	21	8	5	5	7	9	8	10	5	4
Canada	13	20	9	17	15	11	7	11	1	14	14
New Zealand	14	19	7	18	17	21	21	4	9	3	19
Belgium	15	13	16	2	2	10	11	21	20	17	15
Austria	16	12	19	6	6	12	13	16	19	11	8
Australia	17	18	15	19	20	14	8	20	15	1	7
France	18	11	17	16	16	8	5	19	21	16	17
Ireland	19	16	18	3	4	13	20	6	2	20	16
South Korea	20	1	21	7	7	3	6	17	5	4	3
Israel	21	10	5	14	14	19	16	18	7	10	12

Sources: World Economic Forum, Bloomberg News

The world now has a global market based economy (Exhibit 15.2) that connects the market based economies of all nation states through a maze of quasi-governmental economic unions, international trade treaties, and global corporations. The European Economic Union is the most tightly coupled (although unraveling a bit recently) international trade agreement.

EXHIBIT 15.2 GLOBAL ECONOMIC SYSTEM MODEL

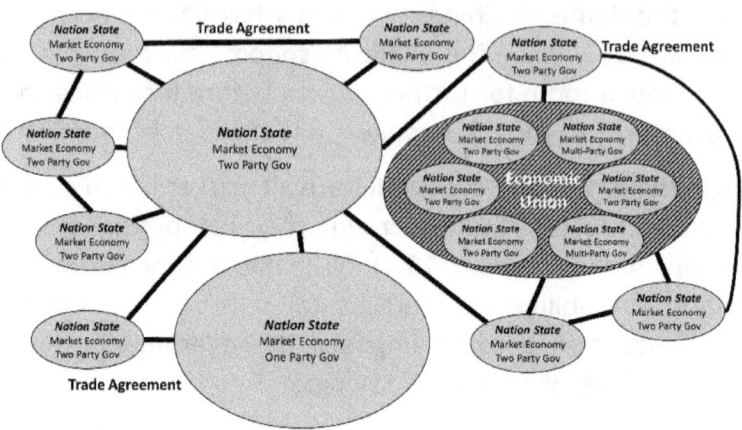

Other regional trade agreements include:

- Andean Community (1969)
- ASEAN–Australia–New Zealand Free Trade Area (AANZFTA) - 2010
- ASEAN Free Trade Area (AFTA) - 1992
- Asia-Pacific Trade Agreement (APTA) - 1975
- Central American Integration System (SICA) - 1993
- Central European Free Trade Agreement (CEFTA) - 1992
- Commonwealth of Independent States Free Trade Area (CISFTA) - 2011
- Common Market for Eastern and Southern Africa (COMESA) - 1994
- G-3 Free Trade Agreement (G-3) - 1995
- Greater Arab Free Trade Area (GAFTA) -1997
- Dominican Republic–Central America Free Trade Agreement (DR-CAFTA) - 2004
- East African Community (EAC) - 2005
- European Economic Area (EEA; European Union–Norway–Iceland–Liechtenstein) - 1994
- European Union Customs Union (EUCU; European Union–Turkey–Monaco–San Marino–Andorra) - 1958

- European Free Trade Association (EFTA) - 1960
- Gulf Cooperation Council (GCC) - 1981
- North American Free Trade Agreement (NAFTA) - 1994
- Pacific Alliance Free Trade Area (PAFTA) - 2012
- South Asian Free Trade Area (SAFTA) - 2004
- Southern African Development Community Free Trade Area (SADCFTA) - 1980
- Southern Common Market (MERCOSUR) - 1991
- Trans-Pacific Partnership (TPP) – 2016

Other regional trade agreements being proposed and/or negotiated include:

- Union of South American Nations (USAN)
- Pacific Island Countries Trade Agreement (PICTA)
- African Free Trade Zone (AFTZ) between SADC, EAC and COMESA
- Arab Maghreb Union (UMA)
- Asia-Pacific Economic Cooperation (APEC)
- Association of Caribbean States (ACS)
- Bolivarian Alternative for the Americas (ALBA)
- Bay of Bengal Initiative for MultiSectoral Technical and Economic Cooperation (BIMSTEC)
- Community of Sahel-Saharan States (CEN-SAD)
- Economic Community of West African States (ECOWAS)[9]
- Economic Partnership Agreements (EU-ACP)
- Euro-Mediterranean free trade area (EU-MEFTA)
- Economic Community of Central African States (ECCAS)
- Free Trade Area of the Americas (FTAA)
- Free Trade Area of the Asia Pacific (FTAAP)
- GUAM Organization for Democracy and Economic Development (GUAM)
- Intergovernmental Authority on Development (IGAD)
- Pacific Agreement on Closer Economic Relations (PACER and PACER Plus)
- People's Trade Treaty of Bolivarian Alternative for the Americas (ALBA)

- Regional Comprehensive Economic Partnership (RCEP) (ASEAN plus 6)
- Shanghai Cooperation Organization (SCO)
- Transatlantic Free Trade Area (TAFTA)
- Tripartite Free Trade Area (T-FTA)
- China–Japan–South Korea Free Trade Agreement

Any agreement with the word "Free" in its title usually includes provisions that reduce or eliminate tariffs between the parties to the agreement before the agreement was negotiated. Since these agreements are negotiated by governmental organizations, they usually have other provisions that might include which entities control or collaborate on regulatory, standards, legal, taxation, and social matters for specific topics.

Investment decisions in this bilateral and multilateral agreement plethora are made by global corporations, governments, trade councils, and treaty organizations. This is relevant to the digital horizon question because of the importance of the flow of investment, the speed of deployment, and the commercialization success of innovation. The economic systems at all levels have different rates of investment flow, deployment speed, barriers to development, and incentives for investment. The global rate is determined by the facilitation or barriers created internally within the nation states and externally through the trading patterns between them.

The two cycles of economic behavior that determine the digital horizon, the Productivity Improvement Cycle and the Innovation Cycle must operate within this global network of trade incentives, barriers, and investment decision making organizations. The good news is that most of the economic systems around the world are market oriented which means that the forces of supply and demand will help guide the global flow of investment to find the best opportunities for a financial return from productivity improvement. The bad news from the perspective of the domestic nation state economies is that this "globalism" of international trade agreements also permits the flow of jobs from one nation of high wage rates to another of low wage rates as businesses seek the lowest cost of economic resource inputs. This trans-border job flow compounds the job destruction issues created by technology.

However, the pendulum may be swinging back because of automation innovations that are reducing the content of low-skilled (and therefore, low hourly rate) labor in many manufactured goods and value adding services. The more activities that are performed by robotics, the less dependency there is on the geography and cost of the labor inputs. An assembly robot applied to a production line in China should cost about the same (leaving aside currency valuations) as the same robot applied to the same type of production line in the US or Europe. As labor content is reduced in key activities, the location and value of supply chains, transportation costs, and the structure of retail distribution networks becomes more important. This trend of "on-shoring" due to technology is just beginning to emerge (Exhibit 15.3), but should pick up speed as automation penetrates deeper into the work flows of more industries.

EXHIBIT 15.3 RESHORING TREND

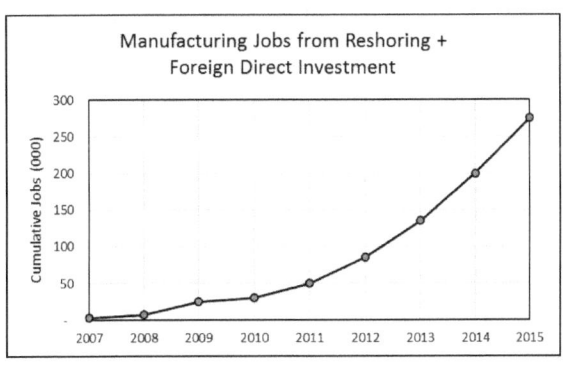

Source: Reshoring Initiative

In short, history is telling us that the economic systems that are the most conducive to economic growth and standard of living improvements are market based (both domestically and globally), innovative, and provide for a financial return to investors.

Chapter 16

How Fragile is the Digital Horizon?

The very nature of digital technology is inherently fragile and always at risk. Any number of threats caused by human ill will or incompetence, natural phenomena, or system design lurk just beneath the shiny brilliance of the digital world. Life is full of risks independent of how much digital technology may be deployed, but digital technologies brings with them new types of risks. While the risks that digital technology brings with it could be viewed as a barrier to the emergence of AI domination the hopeless view of the horizon, they could also be viewed as a barrier to the emergence of the hopeful version. Looking at the various types of threats helps reveal what types of solutions need to be part of the hopeful vision of the digital horizon.

Cyber Security Risks

Evil algorithms can be written just as easily as friendly ones. Unfortunately, this is part of the dark and cynical side of the human condition. Because of the nature and complexity of digital systems, poor cyber security is at the top of the list of fragility factors.

There are several sources of cyber risks for digital systems, which include:

- Malware

  Malware is evil software lurking within a digital system. Computer viruses are an early form of malware and have been a plague (and cause for the creation of an industry) on computer systems since the invention of the personal computer. While once considered primarily the province of mischievous code hackers, malware is now a cyber weapon being used and further developed by nation states, organized crime, political and social activists, and other people or organizations that have chosen this approach to enact some type of damage on their chosen target.

Defensive strategies for the malware risks include preventing the placement of malware in the target system, detecting the presence of malware once it has been placed, detecting the presence of malware once it has been activated, and then the elimination of the malware once it has been detected.

Prevention of malware placement is a challenge because most target systems of value are designed to support high volumes of interaction with users outside of the system. Malware can be designed to piggyback on almost any form of digital interaction. Measures to detect attempted placement of malware include rigorous and multiple levels of user identification, pre-scan procedures before user access is allowed, and user access activity behavior monitoring.

Detecting malware once it has been placed inside a system is extremely difficult because it requires frequent scanning of all stored data and algorithms in the search for code that is not supposed to reside in the system.

Detecting malware once it has been activated requires monitoring the behavior of the target system. Monitoring activity is usually designed to identify when system behavior starts falling outside of desired boundaries. For example, if the email system for a major US movie studio starts sending out digital copies of vaulted unreleased movies to recipients in a small island nation, this could be a tip off that the behavior of the cybersecurity system is abnormal. Of course, this means that "normal" system has to be defined before the monitoring strategy can be created and implemented. As in all defensive security systems, the threat has to be defined and anticipated before an effective security system can be built.

Removing or deactivating malware once it has been detected is also a challenge. If the location of the malware algorithm has been detected, then access to it could be sealed off by the operating system. However, if the malware has been detected because of its behavior, then there are few fixes short of taking the infected system offline and doing a complete memory erase, system reload, and reboot. In corporate networks, if malware is suspected of having penetrated all of the desk top and mobile devices on the network as well as the network servers, then all devices have to be taken off line and purged to ensure eradication of the infecting malware.

- Attack from outsiders

  Attack from outsiders can take many forms in addition to the placement of malware. The most insidious attack from outsiders is the attack not immediately detected. Theft of the information in millions of customer accounts have been announced over the last few years by government agencies, banks, online apps, medical institutions, academic institutions, and online service companies. And most of these theft discoveries took place days, weeks, or months after the thefts occurred. Other types of attacks include denial of service (outsiders preventing a website from servicing its customers by overloading it with fake inquiries), digital hijacking (outsiders taking control over a digital device either to use it for their own purposes or to ransom its use back to the owner), and self-destruction (outsiders causing the attacked system to destroy itself through legitimate commands used in destructive ways). Defense against attacks from outsiders includes the same defensive measures described for malware as well as additional physical security measures.

- Attack from insiders

  A troublesome trend in cybersecurity threats is the increasing role that insiders play in giving access to outsiders or in theft or sabotage. The causes for insiders to become digital evil-doers include bribery, extortion, greed, revenge, or mischief. The defensive strategy against attack from insiders needs to be as strong or stronger as the defensive strategy against outsiders. Digital hygiene for attack from insiders includes both physical, organizational, and digital processes.

- User error

    Mistakes made by users often contribute to cyber security risks but should be totally preventable. If a system has been designed such that it depends heavily on users being trained or having the inclination to follow instructions consistently, then it is inherently vulnerable to user error. This is similar to quality control in a manufacturing process: if a process has been designed in a way that variations from desired actions are not prevented, then variations will inevitably occur according to some statistical distribution. The only way to avoid user error is to design the system to prevent it.

- Service provider failure

    Given that the digital world is networked and interconnected, the supply chain of digital product and service providers represents a major source of cyber security risk. Except for the most elementary of digital devices, every system is composed of apps, software, firmware, and electronics provided by any number of suppliers. Any of these supplied devices could contain malware, spyware, or represent a vulnerability to a digital service.

    Something often assumed safe such as message routing in Internet switches can be compromised by nation states that require the supplier of the electronic switches to include software backdoors. Such software allows the central government to read the addresses and content of messages and even block them as they flow between the switches inside of their national boundaries.

Another risk is the degree of dependence of an organization's operational effectiveness on its digital suppliers. If a military, financial, retail, or entertainment organization is dependent on a 100% real time availability of a network, a data center, or multiple digital access points, then the vulnerability of any of those service providers is a vulnerability to the entire user organization.

The defensive strategies for most of these types of risk includes the design of back-up, alternative, or fail-safe system of providers and procedures just in case there is a failure or disruption at one of the service providers. The defensive strategy for protecting messages transmitted within networks with backdoor snooper software is to transmit messages with some type of coding scheme.

- Physical security

  At numerous points in every system, there is a physical to digital interface which inherently represents a cyber security risk. Whether it be the screen on a smart phone, a locked door to the IT department in an office building, or the data lines feeding into a data center, gaining physical access to a digital system can be easier and more intrusive than hacking the software. The physical access can be achieved through occupant negligence, criminal trespass or through internal sabotage.

- Mobile devices

  Since the advent of the strategy of allowing employees to "bring your own device" (BYOD) for use with the company enterprise systems, there is an added set of cyber security risks. The BYOD strategy is being adopted by many organizations that want to reduce cost or improve the convenience of use for their employees. The company does not have to buy a new cell phone or laptop for every employee, and the employee does not have to keep up with multiple devices for personal and work-related use. While most customer facing digital systems have been designed with at least some degree of concern about outside risks from users, customers, and bad actors, protecting a system from the risks of BYOD requires a different level of threat analysis. Defensive measures calls for system design to prevent user mistakes, sabotage, and theft risks.

- Cloud vulnerability

    The vulnerability represented by cloud computing is at the high end of the service provider risk list. As individuals and organizations continue to put more of their data, operations, and transactions on cloud based services, their exposure to risk depends on the quality of cyber security within their cloud provider. When an individual decides to move all of their passwords from a written list on their desk to their account in a cloud service, they have just taken the cyber security risk from a physical domain over which they had control to a cloud domain over which they have no control. Cloud services often represent a major cost and operational advantage for organizations that do not want or have the resources to build their own internal enterprise systems. Defensive strategies here include shopping for the most secure cloud service provider and having a recovery plan that anticipates the risks and potential failures of a cloud computing environment.

- Legacy systems

    It is likely that any organization older than 10 years old has a legacy computer system somewhere in their operations. The larger and older the organization (e.g. banks, airlines, insurance firms, healthcare systems, retail chains), the higher the likelihood that there will be a legacy system from HP, IBM, or other computer maker with custom designed software running applications critical to the enterprise. The bad news is that if these systems have been integrated into a newer network of systems, they may not be protected by the cyber security provisions built into the newer systems. The good news is that by being older and behind in digital technology, the legacy systems security risks could be simpler and more easily defended with physical security and multi-layer user identification.

The general threat of cyber war is real. Cyber war is asymmetrical (just a few hackers can attack national and global targets). Multiple cyber wars are underway at any time. The entities that are on the attack in cyber warfare are organized, communicate widely among themselves about what offensive strategies are successful, and are often well funded. Cyber warfare is being used by nation states for geopolitical and military purposes, by organized and ad hoc criminal groups for financial purposes, and by political activists and terrorists to further their missions.

Digital cyber security strategy must be focused on system design. It must define the true nature of threats. It must describe and model the threat. It must calculate the likelihood of the threat. It must include the design of back-up, recovery, fail-safe, or fail-soft features into the digital system.

Physical Risk

In addition to the physical to digital interface risks that are inherently part of the cyber security risk, there is the risk of an attack on or loss of the availability of the infrastructure supporting the digital world. Communication satellites, cell phone and microwave transmission towers, fiber cables, data centers, and power grids are all part of the global digital physical infrastructure. The threats to infrastructure include everything from natural disasters (hurricanes, earthquakes, floods, etc.) to electro-mechanical failures (failed transformers taking down a power grid, a truck crashing into a cell tower, a construction team accidentally cutting into a fiber cable, etc.) to physical attack by saboteurs, terrorists, or military organizations (anti-satellite weapons, electromagnetic pulse weapons, weaponized drones, pipe bombs, etc.). In the 1950's when AT&T had a monopoly on telecommunications, many of their line switching centers were designed to withstand a nuclear strike. Today, if an anti-satellite weapon is used to destroy a communication satellite in geosynchronous orbit, much of the debris from the destroyed satellite will stay in the geosynchronous orbit slot making it difficult to replace the lost satellite or to ever reuse that orbit slot.

The best defensive strategies for physical risks include decentralized and distributed designs for power and communication systems as well as increased "hardening" of physical sites and resources.

Cosmic Energy Risks

One of the sources of errors or failures in electronic devices are subatomic particles emanating from outer space. We have all experienced crashes or freezes of our electronic devices whether they be blue screens on our laptops, unresponsive screens on our cell phones, or error messages on our smart thermostats. These failures can be caused by the impact of electrically charged particles generated by cosmic rays that originate outside the solar system rather than manufacturing defects or software bugs.

Cosmic rays traveling through the Earth's atmosphere create secondary subatomic particles (neutrons, muons, pions and alpha particles) that penetrate matter of all types on earth. Although such particles have few detectable short term effects on living organisms, they could be the source of some of the random changes in living organisms that lead to evolutionary effects in the long term. However, there are short term detectable effects on the operation of solid state electronic circuitry that include the changing a bit of stored data from a 1 to a 0 or vice versa. Such a "bit-flip" is called a single-event upset or SEU.

Several electronic failures that are suspected to be the result of a cosmic ray caused SEU have been reported. In 2003 in the town of Schaerbeek, Belgium a bit flip in an electronic voting machine added 4,096 extra votes to one candidate. The error was only detected because it gave the candidate more votes than were possible and it was traced to a single bit flip in the machine's register. In 2008, the avionics system of a Qantas passenger jet flying from Singapore to Perth appeared to suffer from a single-event upset that caused the autopilot to suddenly disengage causing the aircraft dove 690 feet in only 23 seconds. In addition, there have been a number of problems in airline computers – some of which experts feel must have been caused by SEUs – that have resulted in cancellation of hundreds of flights (Bhuva, 2017).

A study of SEU rates for consumer electronic devices published in 2004 ((Wee, 2004) included the following conclusions:

- A simple cell phone with 500 kilobytes of memory should only have one potential error every 28 years.
- A data center like those used by Internet providers with only 25 gigabytes of memory may experience one potential networking error that interrupts their operation every 17 hours.
- A person flying in an airplane at 35,000 feet (where radiation levels are considerably higher than they are at sea level) who is working on a laptop with 500 kilobytes of memory may experience one potential error every five hours.

Additional research has been done to measure the impact of cosmic ray caused SEUs on the evolving semiconductor technology miniaturization. The results shown in Exhibit 16.1 are measured in units called "failures in time" or FIT, which is one failure per transistor in one billion hours of operation. Given that there are more than 3 billion cell phones of various types in the hands of users around the world, FITs can add up quickly. Most electronic components have failure rates measured in 100's and 1,000's of FITs.

EXHIBIT 16.1 SEU FAILURE RATE TRENDS

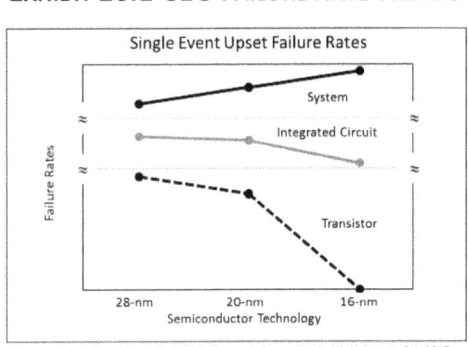

Source: Bharat Bhuva, Radiation Effects Research Group, Vanderbilt University Feb 2017

The trends indicate that although the chances of an SEU decrease as the size of transistors and integrated circuits decrease because they are smaller targets, the density of the packaging of transistors and integrated circuits into systems makes them collectively larger targets. Thus, the chances of an SEU for a system are increasing as semiconductor geometries shrink.

Protecting against SEU's by building shielding around systems is impractical in most cases. A shield of cement or iron many feet thick would be required for the smallest of systems. The only practical protection is to design in redundancy, error checking, and voting circuitry to double check each important operation in real time or as the operations occur. While this is a rational design strategy for space vehicles that travel for years through the cosmos, it seldom is a cost-effective strategy for earthbound products that compete in price sensitive markets. As the market value or importance of earthbound systems increase, design strategies that include protection against naturally occurring phenomena will become economically more practical.

Economic Risks

The risks from cyber security and physical system failures all have threats that can be defined to some level of detail and probability. However, one of the more tangible risks is the economic risk from not enough investment for innovation and technology deployment. Insufficient or diminished investment leads to technological stagnation and then to erosion of the digital world.

Much of the discussion in this book involves the importance and role of the economic factors that lead to the digital horizon. The eventuality of reaching the point of diminishing returns for any new technology is discussed in Chapter 3. Without continual innovation and investment in productivity improving technology, the standard of living stagnates by the limits from diminished returns. These economic risks add to the fragility of the digital horizon because it is the cycle of productivity improvement innovation, investment, deployment that leads to new wealth creation and an improved standard of living.

A Outline For Reducing Digital Fragility

The list of risks and the source of the related threats points to defensive strategies that also contribute to the hopeful vision of the digital horizon. Putting aside the threats of natural and cosmic disasters, the threats from bad human actors can best be defended by system designs that have the following features:

- Idiot-proof (i.e. user mistake proof) interfaces and operations
- No single points of failure (decentralized and distributed system designs)
- Threat-relevant (external and internal) access and operational procedures
- System behavior monitoring and analysis
- Elimination of procedural variation (applying 6-sigma principles)

Each of these features will require further innovation in AI and digital technologies to counter the growing threats from human driven cyber security and physical risks.

With respect to the economic risks, the best defensive strategy is to ensure that the economic policies used by political, regulatory, and financial industry organizations in each country support market-based productivity-improvement investment decision making.

Chapter 17

WILL PEOPLE BE LEFT BEHIND BY AUTOMATION?

Because Of Income Inequality?

The term "income inequality" calls into question whether there is such a thing as its opposite: income "equality". History and economic theory teaches that there is no economic system now and likely never will be that produces income equality (everyone making essentially the same income). By elimination then, "income inequality" is the natural state of every economic system. The more relevant questions are

- Does the socio-economic system provide opportunities for an individual to improve their income?
- Does the socio-economic system lead to an imbalance between the extremes of personal incomes that creates social and political unrest?

The answer to both of these questions are related. The first question is about how much upward mobility a socio-economic system provides.

The answer to question (1) cannot be yes unless there is enough investment to grow the economy, create new jobs at all levels of income, and enough education to support the upward mobility in the different income levels. It also cannot be yes unless the distribution of jobs per income level promotes a growing and sustainable middle class. It also cannot be yes unless social barriers (e.g. caste systems, nobility systems, racial segregation, etc.) to movement between the income levels have been eliminated. Most developed economies have done much over the last century to make the answer yes. The emerging economies have only in recent decades begun removing social and political barriers and promoting more market based economies.

Answering question (2) requires more subjectivity and follows from the answer to (1). If the answer to (1) is yes, then it is likely that there is less chance for social or political turmoil.

If the answer to (1) is no, then there may be barriers to upward income mobility. If the middle economic class is small and the low level economic class is large, then the historical seeds for social and political turmoil have been planted. The economies of early 20th century Mexico, much of South America, and India would have been described by a no answer to (1) and a yes to question (2).

A criticism of the US economy recently has been that there is a growing income inequality problem. A chart used by the camps on both sides of this question is that shown in Exhibit 17.1.

EXHIBIT 17.1 INCOME INEQUALITY

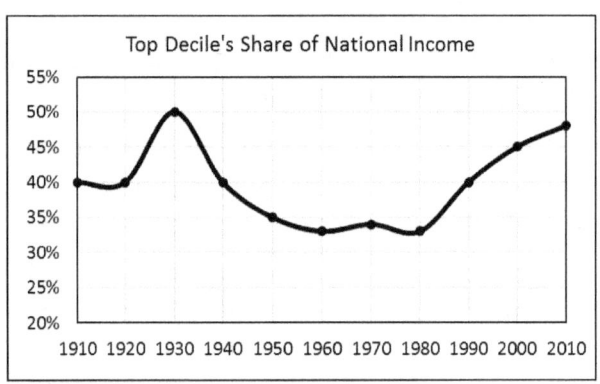

Source: *Capital in the Twenty-First Century*, Thomas Piketty (Harvard University Press, 2014)

This chart shows a growth in the percent of the national income generated by the to 10% of income earners to increase from 33% in 1980 to 48% in 2010. This chart is used to indicate that the top 10% of earners are amassing a larger share of the income being generated by the economy.

Another chart that is used to support the argument that income inequality is extreme is shown in Exhibit 17.2. This chart compares the average income for the bottom 90% of earners by telescoping in on the average income of the top 10%, then 5%, then 1% and finally the top 0.1% of earners.

EXHIBIT 17.2 AVERAGE INCOME COMPARISONS

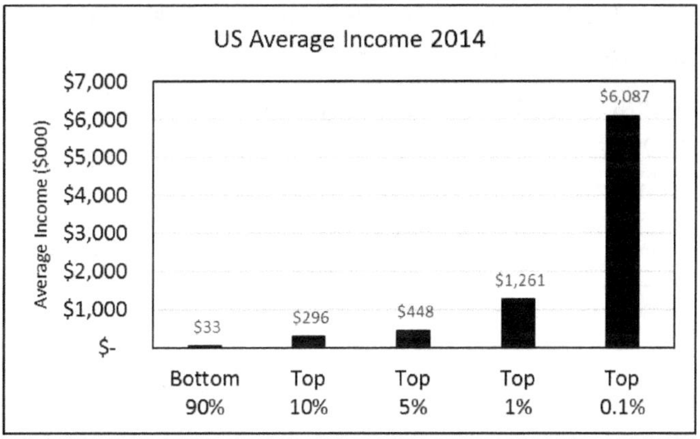

US Average Income 2014

Group	Average Income ($000)
Bottom 90%	$33
Top 10%	$296
Top 5%	$448
Top 1%	$1,261
Top 0.1%	$6,087

Source: Emmanuel Saez, Center for Equitable Growth, June 2015

Both of these charts are designed to support the hypothesis that income inequality is getting worse by showing a trend in Exhibit 17.1 and by implying how extreme the inequality is in Exhibit 17.2. The reality is that the curve shown in Exhibit 17.1 implies that the claim on national income by the top 10% of income earners is cyclical. The percentage was actually highest in 1930 before the Great Depression forced a redistribution of wealth away from the top and while the top marginal tax rates greater than 90% kept it down for many decades. The comparison in Exhibit 17.2 is not much more than an exercise in the statistics of large numbers. The top 0.1% is more than 3 standard deviations from the mean which implies that this chart would probably look similar no matter from which year the data is taken.

What these charts do not show is how much mobility there is between the income groups. The data in Exhibit 17.3 comes from the Internal Revenue Service and provides a better view of income earner mobility. During this recent eight-year period,

- number of taxpayers increased by 28 million (new jobs, better jobs)

- number of people in each income range, except for the lowest, increased (people moving up income ladder)
- number of people in the lowest income range decreased by 6 million (more higher paying jobs than lower paying jobs)
- rate of increase was higher for each higher range (mobility increased the higher up the ladder)

The story that this chart tells is that there were 34 million people (28% of 1996 taxpayers) that saw their income increase significantly over this period, .

EXHIBIT 17.3 UPWARD INCOME MOBILITY OF US TAXPAYERS

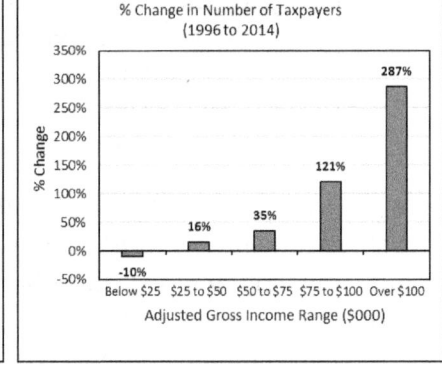

Source: IRS

Exhibit 17.3 also indicates that the number of people that had incomes greater than $100K almost tripled to 24 million. A look at what types of work these people did is shown in Exhibit 17.4. These occupations all require talent, training, initiative, and opportunity. The sources of their income includes wages, income from self-employment, capital gains, interest, and dividends.

So the reality is that there is upward income mobility in the US economy and that mobility can lead to significant improvements in standard of living for a growing number of people. The term income inequality is more a matter of how one measures the distribution of economic success rather than being an insidious flaw in the structure of the economic system itself. When an economy experiences productivity improvement, new opportunities for wealth are created at all levels of income. To participate in this income mobility, a person must have the initiative to get the training needed to develop the talent they have and to seek the opportunities that exist.

EXHIBIT 17.4 OCCUPATIONS AT HIGH INCOME LEVEL

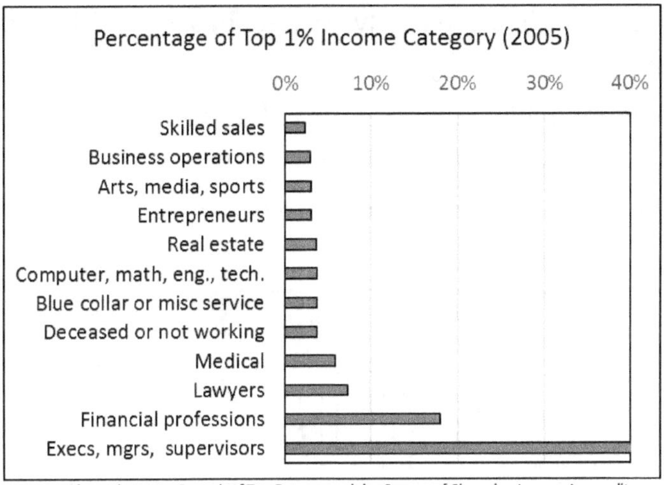

Percentage of Top 1% Income Category (2005)

Occupation	Percentage
Skilled sales	~2%
Business operations	~3%
Arts, media, sports	~3%
Entrepreneurs	~3%
Real estate	~4%
Computer, math, eng., tech.	~4%
Blue collar or misc service	~4%
Deceased or not working	~4%
Medical	~6%
Lawyers	~8%
Financial professions	~20%
Execs, mgrs, supervisors	~40%

Source: *Jobs and Income Growth of Top Earners and the Causes of Changing Income Inequality: Evidence from U.S. Tax Return Data*, Barkija, Cole, Heim, Williams College, 2012

If productivity is improving in the economy, it likely means that obsolete jobs are being destroyed and new jobs that support new business processes are being created. That usually means almost everyone has to consider a job, career, and/or skills change several times during their working life. Those with talent and initiative will find new opportunities to either do as well as they were before or to be able to move up the income ladder. If technology continues to improve productivity, new job creation leads to opportunity creation. For those with low initiative, seeking new training or education to get them qualified to pursue the new opportunities may be a difficult change. For those with low talent, seeking new positions may be difficult. In all circumstances, seeking new jobs may require relocating to new locations. History shows us that there is really nothing new about these challenges; they have faced every generation in modern times.

Because Of Job Destruction?

Technology can destroy existing jobs, create new jobs, change existing jobs, and make jobs mobile. It is not job destruction that could leave people behind but lack of opportunity to deal with the changed jobs, to find the newly created jobs, or to follow the newly mobile jobs.

The simple model in Exhibit 17.5 can be used to describe how and where technology can impact jobs.

EXHIBIT 17.5 JOB CREATION/DESTRUCTION/MOBILITY MODEL

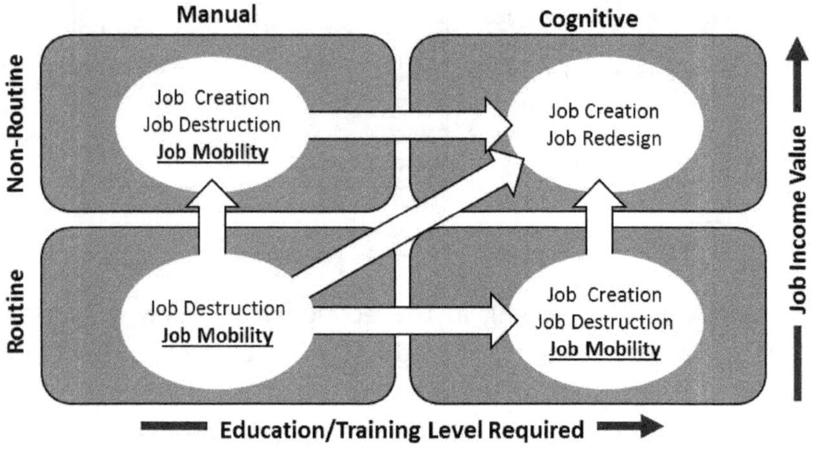

	Manual	Cognitive	
Non-Routine	Job Creation / Job Destruction / **Job Mobility**	Job Creation / Job Redesign	Job Income Value ↑
Routine	Job Destruction / **Job Mobility**	Job Creation / Job Destruction / **Job Mobility**	

◄──── Education/Training Level Required ────►

One of the first types of jobs that became susceptible to mechanical technology was farming. Machines could improve the productivity of the farm worker and provide an attractive return on investment through the replacement of and magnification of human muscle power in manual and repetitive activities. The demand for these farm machines created new jobs in (1) supply chains that were manual and non-routine and (2) in factories that were routine and cognitive. As automation in supply chain companies replaced and amplified the work of people there, it created new jobs in technology and retail companies that were non-routine and cognitive. As factory automation in factories replaced and amplified the work of people there, it created new jobs in automation equipment companies and offices that were non-routine and cognitive. Digital technologies destroy jobs that they replace, create new jobs in industries they help create, change existing jobs that they support, while making many of the jobs more mobile. The term mobile means transferable to lower cost of labor locations.

History shows that automation has tended to create more jobs than are destroyed. The results of a study based on 140 years of census data from the UK (Stewart, De, & Cole, 2015) is shown in Exhibit 17.6. The objective of the report was to test the hypothesis that technology creates more jobs than it destroys. The data shows that while employment in the agriculture industry declined by 95% over the 140 year period, hundreds of thousands of new jobs were created in emerging and entirely new industries such as telecommunications and accounting.

EXHIBIT 17.6 140 YEARS OF UK JOB DATA

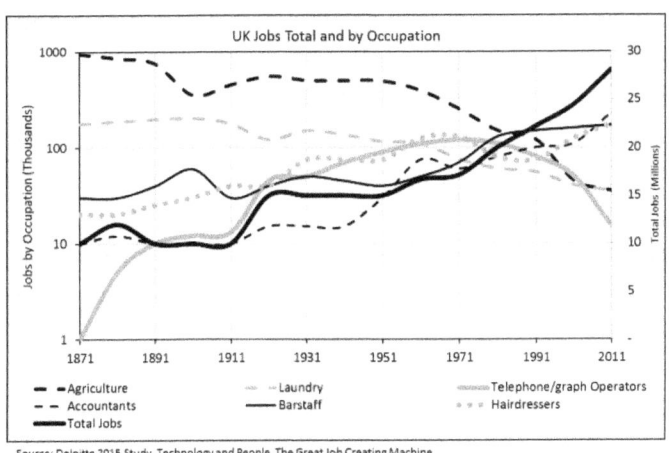

The data also shows that one of the job categories that was created during this period (the telegraph and telephone operators) went through an entire cycle of formation and then destruction. When the telephone and telegraph industry technologies were based on analog electro-mechanical switching, human operators were required to perform the tasks necessary to complete voice calls. Innovation from digital electronics technology took over the manual, cognitive, routing, and non-routine activities in network switching. The job category of telephone operator has almost vanished while the job of telegraph operator was eliminated. New and different jobs were created in engineering, maintenance, sales, and advertising as the role of the human telephone operator disappeared.

While job destruction may get the bulk of the public's attention and job creation is often overlooked, it is the mobility of jobs that often causes many of the socio-political issues facing national economies today. It is unlikely that Henry Ford would have or could have moved the production of the Model-T automobile out of the Detroit, Michigan area in the 1920's. His supply chain, the makers of the machinery he was using on his production line and the producer of the materials he used such as steel and rubber, were all within a few hours of truck or rail transportation in the mid-west states of Indiana, Illinois, Ohio, and Wisconsin. He also had a large supply of educated, trainable, and able bodies to work on his assembly lines.

In 2015, factory automation equipment is made in countries all over the world. Virtually all routine and manual activities have been automated, the education levels in most countries has improved dramatically, and most of the cognitive content of non-routine assembly activities has been taken over by digital technology. The jobs that now exist on the automotive assembly line are globally transportable. The supply chains for parts and materials now stretch around the world. The 2015 version of the Ford Motor Company can shut down entire production lines (Gardner & Snavely, 2016) and move them to lower labor rate and lower tax rate countries. A company can build entirely new production lines in almost any country with the same automation that they would use in the US, and employ a workforce that is just as educated and trainable as they would find in the US counterparts. The company, its stockholders, and investors could then get the economic benefit from productivity improving automation as well as the lower labor costs and lower corporate tax costs.

The tax return data from the previous section and the historical data point to the observation highlighted in Exhibit 17.5 that new jobs created in the cognitive, non-repetitive category tend to require greater education and training (e.g. business, medicine, science, law) and tend to produce greater incomes as illustrated by the arrow pointing upward and to the right in the exhibit. The jobs that have been changed or created by automation in the manual, routine, or cognitive-routine categories tend to require more training, but are also more mobile so that they can be transported to lower labor rate countries. In those cases, job income may be lower even though they may require more training.

As long as there are new innovations in technology that produce productivity improvement, there will be the destruction of job categories and/or the redesign of job content. Simultaneously the new wealth and new opportunities created by technology innovations will create new jobs. There are least three challenges for people facing a disappearing or changing job: (1) get new training for their talents, (2) find the new or alternative job opportunities that match their new training or talents, and (3) increase their mobility to match the mobility of the changed or new job opportunities. These challenges are woven into the fabric of history. These motivations push the waves of immigration and urbanization.

Because Of Education?

The discussion of the above two questions points out the need for education at all levels (basic, graduate, continuing, and job-specific) to be available to those whose jobs have changed due to or have been replaced by technology. This raises several additional questions. Will everyone who needs or wants additional education (1) have access to it, (2) be able to afford it, and (3) have the talent to use it?

Having access to education refers to physical, language, and financial levels of access.

Physical access implies that the source of the instructors, content, and tools is in a location convenient or available to the people that need it. The innovation today in at-distance learning and online learning is beginning to address the physical access challenge.

Digital technology that provides automatic translation, voice-to-text, text-to-voice, natural language processing, and machine learning will contribute to the solution of the language access issues.

The biggest access issue is and likely will always be financial. There are several important questions that have to be answered that are part of the financial access issue.

Who will pay for the education? What should the role be of personal funds, government stipends, or charitable foundations?

How does the person who needs the education accepts responsibility for paying for the education? Will it be in the form of long term loans or future income taxes?

What type of qualification or testing for the education must the person demonstrate? How does the person demonstrate a match between their talent level and the level of education they seek?

Affordability of the needed education is also a big challenge. As discussed in Chapter 11, the education industry is one of the industries in most need of productivity improvement. The cost of education in the US, by almost any measure, has continued to grow at a faster rate than inflation and faster than the growth of the economy for decades. Without a major breakthrough in productivity improvement in the education industry, two dangerous and related trends will continue to pick up momentum. One trend is that education at almost any level will be available only to upper income levels of the population. The second trend is that the debt burden on the newly educated and the tax burden on everyone will consume larger percentages of income. If the affordability of education does not dramatically improve, the risk of people being left behind due to lack of education increases with time.

The question of talent level for each person as a factor of what type of education they will benefit from and what job types they will be able to master is also important. Talent can mean many things. Physical dexterity, abstract thinking, problem solving, memory, mathematical ability, artistic skills, hand-eye coordination, learning speed, social skills, collaboration skills, honesty, and attitude are just a few of the talents that people can develop to one degree or another. As more of the jobs created by technology fall into service industry categories, the breath of skills needed will likely increase. People fall into many different talent levels; not everyone is equipped to be a heart surgeon, and not everyone has the talents to repair plumbing. Matching people's talents with the education needed to perform new jobs will continue to grow in importance as new jobs are created and as old jobs are destroyed or changed.

Who Does Get Left Behind?

Who wins or who stays ahead of the economic parade as technology innovations march in and improve productivity? The winners will be most everyone with some combination of talent, training, initiative, and financial support. Most everyone has a talent for something. The challenge is to get the training needed to develop that talent into something that matches the opportunities that are created. Initiative is needed to pursue the needed training and the new job opportunities.   Initiative often overcomes any imbalance in the combination of talent, training, and financial support.

Who gets left behind or struggles to keep the parade insight? Those who lack the education and training, the initiative to get the right training, the initiative to seek the new opportunities, or the financial support to pay for the education and training will inevitably fall behind.

Will people with talents less than that required by newly created jobs be left behind independent of whether there is enough training or financial support to help match the talents with opportunities? Is technology only creating new jobs that require the human skills that are only available in a portion of the population that is diminishing in size over time?   Is the promise of digital technology based, productivity improving, automation, AI, and robotics creating a growing population of untrainable economic and social non-contributor people?

The answer to these questions is embedded in the answer to the question of how digital technology is designed, developed, and deployed to serve the economic system. People are the economic system and the technology that supports and fuels the economic system is designed by people. As long as there are two people that want to have dinner together, a fully automated restaurant that is devoid of human interaction will not be viewed as appealing. As long as people want to improve their lives economically, academically, or socially, they will not surrender their lives to machine intelligence but will design it to serve them. A big part of the digital horizon is how we innovate and apply technology to serve the needs of a population that has a diversity of skills, educational levels, and degrees of initiative.

Chapter 18

IS INNOVATION NEEDED IN SOCIAL SERVICES?

The Short Answer: Yes!

The economies of Europe and North America have over 80 years of experience with various types of welfare systems that have consistently grown to consume larger portions of the wealth in each country. The great communist experiments in totalitarian socialism in the former USSR and China all fell far short of their declared goals of leveling out and distributing wealth across entire populations. Both of these socio-economic-political systems collapsed with the replacements being market economies that included some type of democracy. Even in these relatively new market based economies, the cost of social services has grown consistently faster than the economies.

The cost of government funded social welfare and human services continues to grow as a percent of the GDP and is at historical highs for almost every developed country (Exhibit 18.1). The countries with the highest social expenditure as a % of GDP tend to be in the central and northern European countries where the populations have demanded more social support by their governments. These countries, like many developed countries, are experiencing declining birth rates and aging populations.

EXHIBIT 18.1 SOCIAL SPEND AS % OF OECD GDP

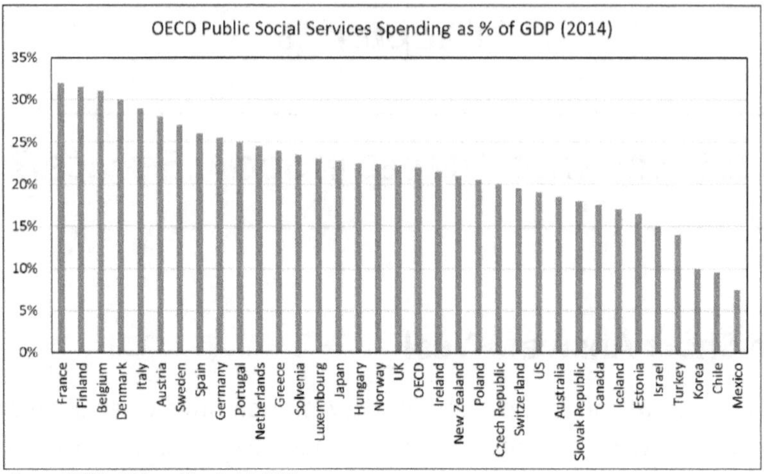

OECD Public Social Services Spending as % of GDP (2014)

Countries (left to right): France, Finland, Belgium, Denmark, Italy, Austria, Sweden, Spain, Germany, Portugal, Netherlands, Greece, Slovenia, Luxembourg, Japan, Hungary, Norway, UK, OECD, Ireland, New Zealand, Poland, Czech Republic, Switzerland, US, Australia, Slovak Republic, Canada, Iceland, Estonia, Israel, Turkey, Korea, Chile, Mexico

Source: OECD Social Expenditure Update 2014

As these high levels of social expenditures continue to rise, there have been economic and political factors that are making several countries question the level that they can or should afford. The inability of Greece to finance its budget deficits has resulted in the cut back of social services spending. Italy and Spain are approaching similar situations as Greece. Scandinavian countries are beginning to question how to increase the participation of their populations in their work forces for the lower and higher end of the age spectrum. Denmark has already increased its retirement age and is now considering reducing the time that students are eligible for educational grants to get them to graduate and into the workforce earlier (Levring, 2017).

While productivity improvement and innovation have been terms and concepts developed in the manufacturing industries for centuries, and in services industries for a few decades, these terms have not been associated with the social services until recently (Rønning, 2015).

Rønning and Knutagard make the case that innovation is imperative. In their words: "Both service users and employees within social welfare and human services are facing the demand for more efficiency, and the consequences of efforts to realize it. Employees can feel inadequate even when they stretch the rubber band too far. Into this context, innovation is introduced as an imperative to be even more efficient – one more demand." Their concern is that productivity improvement innovations will be made by people outside of the industry and will be imposed on the workers in the industry rather than in collaboration with the industry.

The concern about the imposition of an outside solution is well founded and calls for the application of the best practices learned from decades of innovation in the manufacturing and technology industries. Any innovations must be made from the perspective of business processes. The people who understand the strengths and weaknesses of any business process are the people working in it. For any innovations to be made to social services, there must be a collaboration between the stakeholders (within the process and served by the process) as well as people who may have solution tools to provide the innovations.

Innovations are needed in how the request for services are validated, how the services are delivered and monitored, and how they are financed. The problems that exist today for social and human services include

- Duplicated and wasted resources due to overlapping services from different organizations

    Possible solutions could include (1) an integration and streamlining of the different organizations and networks of contractors and suppliers that administer and deliver services, (2) automation of key manual and cognitive routine activities, (3) smart sensors and home automation technology to monitor the status of people receiving services, and (4) data integration and data analytics to optimize the effectiveness of the services being delivered. AI systems, home automation, wearable electronics, and data analytics are just some of the technologies that could be used to eliminate waste and duplication while improving quality.

- Fraud due to inadequate validation and feedback from service requestors and/or providers

    Technologies that are being developed to deal with consumer credit fraud, identity theft, and cybersecurity would be applicable to this problem

- Errors and mistakes due to unreliable and highly variable service processes

    Reengineering the service delivery processes and applying automation where it is cost effective to eliminate the sources of variation and mistakes will solve these problems. There are many solutions and lessons learned from manufacturing and consumer service industries that would be directly applicable.

- Standardized rigid service structures that under serve or over serve individual needs

    Reengineering the service delivery processes would create the opportunity to build in flexibility in the process to optimize the combination and delivery of services needed to satisfy user needs.

- Incentive systems for service providers to improve and for service receivers to use no more than needed or to reduce or eliminate their need.

    Today there are few policy or operational mechanisms in place to incentivize improvements on the part of the user to reduce or eliminate their needs. Bringing lessons learned from market based services might provide insights into how to innovate in this area.

While it is intuitive that much of the automation technology that has been used to improve productivity in the production and professional services sectors could be used to foster innovation in the social and human services sector, design and implementation of such solutions face serious barriers. Changes designed to improve the productivity and reduce the cost of social and human services would likely require changes to government organizations and policies, redefinition of service goals and objectives, changes to job definitions and work rules, as well as the investment in and implementation of a variety of technologies. This will likely be a slow process but one which is essential.

Is Basic Income A Solution?

The term "Basic Income" was recently coined to describe an old idea: Pay everyone a minimum income whether they are working or not. This simplistic concept is not all that different from an "allowance" that a parent may give their child for spending money. This concept is a recycled idea left over from the failed communist experiments in central government planning. It is also at the heart of every social welfare system in existence today.

In the US it started with the Social Security System in the 1930s and expanded with numerous welfare programs in the following decades. In post-WWII Europe, virtually every nation established social welfare programs that were aimed at providing a minimum level of income and services for the non-working and low incoming earning portions of their populations. In the US, this has also been called establishing a safety net for those people that either fell off of or who never got on the income growth ladder.

Most currently implemented welfare systems or social safety nets have complicated rules and time horizons for who is qualified to receive government payments. Large bureaucratic organizations were created to administer and manage a wide variety of government payment programs.

The intellectual appeal of the concept of a guaranteed income level includes:

- an easily defined way to redistribute wealth
- a simple system that would eliminate the need for the large bureaucracies to manage the redistribution
- a built-in stimulate for the economy

- a way to deal with all the lost jobs taken by robots (including the former workers eliminated in the implementation of a Basic Income system).

Opponents to the concept argue that there are more questions created than answers provided and many severe negatives that would off-set any benefits. Questions such as:

- What level of basic income is meaning or helpful?
- What level of basic income is "fair"?
- Would not the simplistic egalitarian nature of it imply that basic income is a basic human right creating a large new set of legal, political, and moral issues?
- Would not the work incentive be eliminated in portions of the population that might need it the most?
- Would the self-improvement incentive be eliminated for the people that would want to get the education and training needed to work in new jobs created by automation or available elsewhere in the economy?
- How would a cap be put on an ever growing need for new taxes?

  The cost of a basic income system can be enormous. An economy the size of the US could pay all citizens a basic income of about $10,000 a year if it doubled the average tax rate to match that of Denmark and eliminated all other social programs including Welfare, Social Security, and Medicare. That would be a social upheaval turning over 80 years of social services that would not be politically feasible even if a rational basis for it could be established.

- How much of the capital needed for investment in technology that improves productivity would be consumed?
- Would productivity actually decline (as it has over the last several years)?
- Would tensions increase between those who are net payers of money to the government (i.e. tax payers) and those who

are net recipients of money from the government (i.e. non-tax payers). A universal basic income would simplify, clarify, and focus that tension between those who opt out of the work force and those that do not.
- If basic income is not considered as a universal replacement for all existing government welfare systems, is it not just another line item in welfare state budgets?

The basic income concept does not address what could be some of the biggest psychological issues that may be at the core of our future. If a person does not have any reason or purpose to get up in the morning, why should they? What will people do with their lives if the state has taken over their financial responsibilities?

Basic income is just one possible answer to a problem that does not exist now and may never exist: the end of employment due to technology. Historical evidence, recent experience, and thoughtful analysis (some of which is provided herein) all point to a human guided evolution of a symbiosis between people and machines. Before governments begin planning for an automated world without work, innovation in the social services that exist today would be a more rational strategy.

Chapter 19

WILL AUTOMATION SOLVE ALL PROBLEMS?

The Short Answer: No!

Unless there is a sudden surge in the evolution of humans to be more caring, less war-like, more peaceful, more ethical, more tolerant, more cooperative, and more philosophical, automation cannot not solve all the problems of humanity. It is just as much science fiction to assume that automation will solve all problems for all people as it is to assume that technology will grow into an AI that will take over the world and all the people in it.

Despite More Abundance?

The most radical optimistic vision about technology is that it will improve productivity in agriculture, manufacturing, and services to the point that the law of supply and demand will be repealed. This vision has it that there will be enough food so that no one goes hungry, enough materials so that no one will be in need of "things", enough healthcare so that no one will be sick, and enough information so that all decisions will be simple. This is a wonderful vision. If one could suspend reality from their thinking, one could plot out trends in technology that could be extrapolated far into the future to imply that all this might happen. Besides the limitations that reality puts on this vision, there is one other problem. What is the definition of enough?

One of the key points of Maslow's hierarchy is that as people have their basic needs satisfied, they develop new needs at a higher level in his hierarchy. In fact, this tends to be the driving force in a growing economy. As more wealth is created, certain lower level needs are satisfied. The demand for higher level needs increases such as luxury goods, higher priced food, longer vacations, etc. Abundance at one level of need often creates shortages at higher levels because of more demand. At this point the thought that technology would produce an abundance that solves all human problems is as much science fiction as the fear that technology will replace all jobs.

Because Of More Decisions?

As technology advances with new discoveries in electronics, materials, and biosciences, the decisions that people will have to make could become more profound. If synthetic genetics could give parents the ability to select the human features for their next baby much like picking items from a menu, a plethora of new issues would then have to be considered the yet to be conceived child. If the length of a person's life could be engineered by making changes in their genetic code, then a new set of decision problems will be created. If a person could hand over to a home automation AI system all decisions about the occupant's life style, new self-identity problems might be created.

All of these questions are hypothetical but point out that fact that as each new wave of technology crashes into the economy and solve old problems, new issues are created and new types of decisions must be made. Each new wave of technology will bring along answers to old questions, but will create new questions that will call for new innovations..

Even If It Did, Why Get Up In The Morning?

Usually the first order of important business when one gets up in the morning is determining what to do that day. For those that are gainfully employed, it means getting dressed and getting to the location where their paid duties for the day will be performed. For those that are stay at home parents, it means managing and planning for the wellbeing of children. For those in retirement, it means getting dressed and getting to the location where they can be useful (visiting friends and family, volunteering) or self-fulfilled (playing golf, attending a book club meeting). For those that are school age, it means getting dressed and getting to the location where they will listen, study, and learn something that might be helpful someday.

But what if there was no need to go to work or school? What if automation and AI has been developed and implemented to the point where all of our needs are being met by a machine. For the sake of this thought puzzle, the question of whether the machine is owned by the government or a corporation is not a concern. Nor is how much income everyone gets or where it comes from. Just assume that when it time to wake up every morning, there are no physical issues or challenges. Assume that there is no reason to want to be better. The big question then is why get up in the morning?

While this thought puzzle is hypothetical, the questions asked are not all that different from what a ward of the state (e.g. prison inmates) faces each morning. So even though technology is making improvements in the economy, creating more wealth, and making quality of life better, it is not helping answer the question: Why get up in the morning? That will remain a uniquely human decision. And the need to answer this question is one of the underlying human traits that will make people want to stay in control of the technologies they innovate.

SECTION VII

CONCLUSIONS

Chapter 20

DIGITAL HORIZON IS HOPEFUL

The digital horizon is where we are headed with the help of technologies that are based on the simple activity of adding binary digits. Digital technology goes by many names. A short list includes: robotics, automation, artificial intelligence (AI), machine intelligence, big data analytics, smart sensors, and digital assistants. Whatever the name, its solid state heart beats with pulsing electricity that feeds an integrated circuit of nanoscale calculators.

Fear of technology started some time before the Luddites. It persists today at an amped-up level exceeded only by the most extreme science fiction. Hardly a week goes by without there being a headline spawned by a quasi-credible source that some new technology has added new evidence to the pile that our robotic technology overlords are just around the corner. It used to be that blood curdling crime headlines sold newspapers. Now that we are immersed in 24 by 7 digital media, the stakes of capturing eyeballs seem to require a never-ending stream of headlines that reinforce the fear of an apocalyptic end to our species. If it isn't the weather, some incurable face eating disease, or a wayward comet that will do us in, then it is technology. The doomsday technology headlines often imply that it is precisely technology that is the cause of all the other global threats.

On the other hand, there are often breathless headlines touting the wonders of an invention or discovery that has the promise of eradicating a major illness, solving a current social problem, or creating an abundance of everything so that all needs are met and all worries diminished. Such is the world in which we live. The reality is the future that technology is bringing us is hopeful but likely something short of pure abundance (Exhibit 20.1).

EXHIBIT 20.1 DIGITAL HORIZON METER POINTS TO HOPEFUL

There are no more reasons today for us to expect that technology will lead to our end than there were hundreds of years ago. There are three reasons that the digital horizon meter tilts toward the hopeful side rather than the hopeless.

The first reason is **economics**.

Humanity's drive to be better is an innate attribute and one of our unique features. Improving our standard of living, growing more and better food, making more and stylish clothing, achieving better healthcare, increasing our education are just of a few of the motivations that guide our activities globally. Once we learned that we could work together collaboratively in an economic system, we found that innovation of tools, techniques, and technology was the secret to producing more with less. Improving economic productivity led to improving the standard of living.

Throughout history it has been shown that for a new technology to be accepted by society, it must have a beneficial economic impact. For it to have a beneficial economic, it must improve productivity. When productivity is improved, new wealth is created. When new wealth is created, quality of life improves. For new technology to improve productivity, it must be deployed in the economy. For it to be deployed in the economy, there must be investments made. If there is no return on investment (ROI), then no new technology will be developed and deployed.

In short, if no ROI, then no AI!

The second reason is **human intelligence**.

Humans are inventing, developing, and deploying technology. People are creating artificial intelligence essentially as mirror images of human intelligence. People are using tools invented by people such as algorithms, neural network structures, and data analytics to build intelligent technologies. People are teaching it with data from sensors and libraries of all human knowledge.

Digital technology can do no better than use mathematical approximations of how the universe operates. Technology can operate physical tools and execute information tools many times faster than human speed. However, the reality is that the artificial version of intelligence is faced with the same challenges as the human version. Decisions have to be made with imperfect data collected from an uncertain world. Using digital technology, we are making faster, more productive, electronic versions of our imperfect selves.

People design machines to be helpful, to improve productivity, to collaborate on decisions. It is all part of the human trait to be better. The only way for technology to become humanity's overlord would be by the decision of people. People would have to design machines that do all the thinking, make all the decisions, do all the work, and take care of all human needs. People would have to make the investments to have all the necessary machines built. And then people would have to surrender to all the machines.

In short, human intelligence controls artificial intelligence by design.

The third reason is **necessity**.

Most of the developed economies have populations that are aging. Birth rates have dropped and life spans have increased to the point where the average age in many countries is getting older every year. The ratio of the portion of the population that needs support (preadult and post-retirement) to the portion that is working gets larger every year. To offset and diminish the chances of a severely diminished quality of life due to the aging population, there is a need for more, not less, technology to sustain and accelerate productivity growth.

Another factor that drives the necessity of more technology innovation is the reality of the law of diminishing returns. As any one new productivity improving innovation (a machine, a software package, a new methodology) is deployed throughout an economy, at some point it's benefit diminishes to nil. No additional unit of input from this innovation will produce an additional unit of output. Productivity does not grow. It flattens out. If the population keeps growing or aging, productivity improvement becomes negative and productivity declines. Less is produced by more.

In short, society needs more technology not less.

Integral to each of these three reasons is the critical factor of improving productivity. Improving productivity simply means doing more with less in an activity, process, or system. A process or system is one or more activities being performed by a person or machine. Improving the productivity of a process means achieving more output with the same or less input. Output is anything that can be measured. Input can be time, materials, or labor hours.

Just throwing technology at the problem of improving productivity in a process will not work unless several process design rules are followed.

The first rule is to design the workflow of activities in a process to take advantage of the capabilities of the human and machine resources available. If the task at hand is to improve the productivity of an existing process, then all activities in the existing process have to be identified as to whether they add value or not. Then redesign the workflow to eliminate as many of the non-value adding activities as possible. After workflow redesign, each activity has to be evaluated as to whether it can be automated. The cost of automating or not automating each activity has to be estimated. The benefits of automating or not automating each activity has to be estimated. If necessary, going through this cycle of analysis might be helpful. Of course, at each step of the analysis, tradeoffs will have to be made as to what combination of person and machine will provide the best cost performance benefit. After all the analysis is done, a final process design will emerge with designations on which activities will be performed by machines alone, by people alone, and by people working with machines. If the task at hand is to create a new process, then the initial step is to create the set of activities to be performed in the new process and then go the same analysis outlined above.

The second rule is to eliminate unwanted variations in each activity. Variations cause mistakes, mistakes cause quality problems, quality problems cause extra cost. Variations can be eliminated by (1) replacing human activities that are susceptible to errors with machines that are not, (2) creating tools, fixtures, or other aids that eliminate the sources of variation by a human or machine, and (3) redesigning an activity to eliminate the sources of variation.

A third rule is to include flexibility in a process if it expands the value and usefulness of the process. An automobile production line that only produces black two door four passenger sedans will have less value than a line than can produce any model sedan with any color and any seating arrangement in the product catalogue. A hamburger shop that only makes one style of hamburger will likely have fewer customers than a shop that can make a variety of combinations. Automation, especially with information intensive technology, can be a key part of designing a process that has a wide variety of cost effective flexibility.

A fourth rule is to identify the role and value of information in a process. Automating the capture and analysis of data that supports a business process or is part of the output is often part of the innovation in the process. Designing the data capture, flow, analysis, and conversion to information is a critical part of any process design.

A fifth rule is to identify how much decentralization and autonomy is necessary or beneficial in the process design. Making sure the order entry kiosk in a hamburger restaurant is separate from the power supply of the fryer is likely a lower priority than having back-up generators for a hospital operating room.

When an old process has been reengineered or a new process has been designed, typically a mixture of people and machines is the best answer. The use of robot arms in a car painting booth on an assembly line eliminated all the human labor that had formerly handled the spray nozzles. But new jobs were created for people to program, fix repair, supply, and setup the robots, and to evaluate the painting quality data from the sensors on the robot system.

In summary, the three reasons above will prevent humanity from being overmastered by its own creation. History has shown that there is a symbiotic relationship between technology and society. Technology is changing how we behave, how we learn, and how we adapt biologically while we make technology work to further society's needs and desires. The flow of information and activities in all processes will repetitively be evaluated and redesigned to be better. Activities will be automated where beneficial. Job content will be changed to reflect the need for new combinations of human skills. New wealth created can be used to give people the opportunity to get new education, to be retrained, or to develop new skills. How chaotic or how graceful the transitions happen will be determined in large part by the flexibility of existing social and governmental institutions as well as by the degree of democracy in the economic systems.

While there remain unknowns in the journey, all indications are that humankind and technology will evolve together while sailing toward the digital horizon.

References

Acemoglu, D., & Robinson, J. (2012). *Why Nations Fail: The Origins of Power, Prosperity, and Poverty.* New York: Crown Publishers.

Albergotti, R. (2014, March 24). *Zuckerberg, Musk Invest in Artificial-Intelligence Company.* Retrieved from Wall Street Journal: http://blogs.wsj.com/digits/2014/03/21/zuckerberg-musk-invest-in-artificial-intelligence-company-vicarious/

Assimov, I. (1942, March). Runaround. *Astounding Science Fiction.*

Bellman, R. E. (1961). *Adaptive Control Processes: A Guided Tour.* New York: Princeton University Press.

Bhuva, B. (2017, Feb 17). *Particles from outer space are wreaking low-grade havoc on personal electronics.* Retrieved from Physics.Org: https://phys.org/news/2017-02-particles-outer-space-wreaking-low-grade.html

Bjorkland, D. F. (2007). *Why Youth Is Not Wasted On The Young: Immaturity In Human Development.* Oxford: Blackwell Publishing.

Brustein, J. (2016, August 15). *Uber and Lyft Want to Replace Public Buses.* Retrieved from Bloomberg News: https://www.bloomberg.com/news/articles/2016-08-15/uber-and-lyft-want-to-replace-public-buses

Burger, D. (2016, October 27). *Goldman's Multifactor Robots: A Post-Human Investing Guide.* Retrieved from Bloomberg News: https://www.bloomberg.com/news/articles/2016-10-27/goldman-s-multifactor-robots-a-post-human-guide-to-investing?adv=avishares

Carey, K. (2014, June 28). *The Upshot.* Retrieved from New York TImes: https://www.nytimes.com/2014/06/29/upshot/americans-think-we-have-the-worlds-best-colleges-we-dont.html?_r=0

Cohen, J. (Performer). (2017, February 20). *Ringling College of Art + Design Town Hall Series.* Van Wezel, Sarasota, FL, USA.

Encyclopaedia Britannica. (2016). *Diminishing Returns*. Retrieved from Britannica.com: https://www.britannica.com/topic/diminishing-returns

EU Parliament Committee on Legal Affairs. (2016, May 31). *Draft Report with Recommendations to the Commission on Civil Law Rules on Robotics.* Retrieved from European Parliament: http://www.europarl.europa.eu/sides/getDoc.do?pubRef=-//EP//NONSGML%2BCOMPARL%2BPE-582.443%2B01%2BDOC%2BPDF%2BV0//EN

Fagnant, D., Kockelman, K., & Bansal, P. (2015). Operations of a Shared Autonomous Vehicle Fleet. *Transportation Research Record, No. 2536*, 98-106.

Federal Bank of Dallas. (2016, 5 5). *www.dallasfed.org/educate/everyday/capital.* Retrieved from www.dallasfed.org: http://www.dallasfed.org/educate/everyday/capital

Finley, K. (2012, September 12). *Did a Computer Bug Help Deep Blue Beat Kasparov?* Retrieved from Wired: http://www.wired.com/2012/09/deep-blue-computer-bug/

Flynn, J. R. (2007). *What is Intelligence?* Cambridge University Press.

Forrest, C. (2014, March 5). *Google and robots: The real reasons behind the shopping spree.* Retrieved from TechRepublic: http://www.techrepublic.com/article/google-and-robots-the-real-reasons-behind-the-shopping-spree/

Gardner, G., & Snavely, B. (2016, September 16). *Ford shifting all U.S. small-car production to Mexico.* Retrieved from Detroit Free Press: http://www.freep.com/story/money/cars/ford/2016/09/14/mexico-ford-shiftng-us-car-production-mexico/90355146/

Gilbert, D. (2016, January 20). *Davos 2016: Need To Embrace Robot Revolution Not Fear It, Tech Leaders.* Retrieved from International Business Times:

http://www.ibtimes.com/davos2016needembracerobotrevolutionnotfearittechleaderssay2272199

Harari, Y. N. (2014). *Sapiens: A Brief History of Humankind.* . Harper.

Hodges, A. (1983). *Alan Turing: The Enigma.* Princeton and Oxford: Princeton University Press.

Iowa State University. (2016, 8). *Grain Harvesting Equipment and Labor in Iowa.* Retrieved from https://www.extension.iastate.edu/agdm/crops/html/a3-16.html

Jason Richwine, P. (2012). *Government Employees Work Less Than Private-Sector Employees.* Washington, D.C.: The Heritage Foundation. Retrieved February 24, 2017, from http://thf_media.s3.amazonaws.com/2012/pdf/b2724.pdf

Johnson, B. (2015, July 5). *Big Spenders on a Budget: What the Top 200 U.S. Advertisers Are Doing to Spend Smarter.* Retrieved from Advertising Age: http://adage.com/article/advertising/big-spenders-facts-stats-top-200-u-s-advertisers/299270/

Joy, B. (2000). Why the Future Doesn't Need Us. *Wired.*

Kambayashi, S. (2014, January 18). The Onrushing Wave - The Future of Jobs. *The Economist.*

Keynes, J. M. (1963). Economic Possibilities for our Grandchildren. In J. M. Keynes, *Essays in Persuasion* (pp. 358-373). New York: W. W. Norton & Co.

Kumagai, J. (2016, December 30). *The Grand Ethiopian Renaissance Dam Gets Set to Open.* Retrieved from IEEE Spectrum: http://spectrum.ieee.org/energy/policy/the-grand-ethiopian-renaissance-dam-gets-set-to-open

Levring, P. (2017, March 1). *Welfare Icon Now Wants People to Take Care of Themselves.* Retrieved from Bloomberg News: https://www.bloomberg.com/news/articles/20170301

Lewis, J. (2016, September 6). *Robot Macroeconomics: What can theory and several centuries of economic history teach us?* Retrieved from Bank Underground:

https://bankunderground.co.uk/2016/09/06/robot-macroeconomics-what-can-theory-and-several-centuries-of-economic-history-teach-us/

Lewis, M. (2003). *Moneyball: The Art of Winning an Unfair Game.* New York: W. W. Norton & Company.

Lewis, M. (2014). *Flash Boys: A Wall Street Revolt.* New York: W. W. Norton & Company, Inc.

LII of Cornell University Law School. (n.d.). *Legal Information Institute.* Retrieved February 18, 2017, from Cornell University Law School: https://www.law.cornell.edu/wex/twinkie_defense

Lovgren, S. (2005, 8 31). *Chimps, Humans 96 Percent the Same, Gene Study Finds.* Retrieved from National Geographic: http://news.nationalgeographic.com/news/2005/08/0831_050831_chimp_genes.html

Maestas, N., Mullen, K., & Powell, D. (2016). *The Effect of Population Aging on Economic Growth, the Labor Force, and Productivity.* Washington, DC: National Bureau of Economic Research.

Mark, G. A. (1987). The Personification of the Business Corporation in American Law. *University of Chicago Law Review*, Volume 54, Issue 4, Article 9.

Markoff, J. (2011, February 16). *Computer Wins on 'Jeopardy!': Trivial, It's Not.* Retrieved from New York Times: http://www.nytimes.com/2011/02/17/science/17jeopardy-watson.html?pagewanted=all&_r=0

Markoff, J. (2011, February 17). Computer Wins on 'Jeopardy!': Trivial, It's Not. *New York Times.*

Maslow, A. (1954). *Motivation and personality.* New York: Harper.

Mokyr, J., Vickers, C., & Ziebarth, N. (2015). The History of Technological Anxiety and the Future of Economic Growth: Is This Time Different? *Journal of Economic Perspectives—Volume 29, Number 3*, 31-50.

Moore, G. E. (1965, April 19). Cramming More Components Onto Integrated Circuits. *Electronics, 38*(8).

Morgenstern, M. (2016, June 25). Automation and anxiety: Will smarter machines cause mass unemployment? *The Economist.*

National Human Genome Research Institute. (2010). *2010 National DNA Day Online Chatroom Transcript.* Retrieved from National Human Genome Research Institute: https://www.genome.gov/dnaday/q.cfm?aid=785&year=2010

Peter Diamondis, S. K. (2012). *Abundance: The Future Is Better Than You Think.* New York: Free Press.

Price, E. (2015, December 23). *Elon Musk Nominated for Luddite of the Year.* Retrieved from The Guardian: www.theguardian.com/technology

PwC. (2016). *The MoneyTree Report.* New York: PwC.

Revolution, A. H. (2016). *The Triple Revolution.* Retrieved from Published Papers and Official Documents: http://scarc.library.oregonstate.edu/coll/pauling/peace/papers/1964p.7-03.html

Ricardo, D. (1812). *On the Principles of Political Economy and Taxation.* London: John Murray.

Rønning, R. K. (2015). *Innovation in Social Welfare and Human Services.* New York: Routledge.

Schwab, K. (2016). *The Fourth Industrial Revolution.* World Economic Forum.

Shakelton, R. (2013). *Total Factor Productivity Growth in Historical Perspective.* Washington, DC: Congressional Budget Office.

Shanghai Jiao Tong University. (2015). *Academic Ranking of World Universities.* Retrieved from Shanghai Ranking: http://www.shanghairanking.com/ARWU2015.html

Silberman, S. (2009, August 24). *Placebos Are Getting More Effective. Drugmakers Are Desparate To Know Why.* Retrieved from Wired: https://www.wired.com/2009/08/ff-placebo-effect/

Smith, P. (2015, March 23). *Apple co-founder Steve Wozniak on the Apple Watch, electric cars and the surpassing of humanity.* Retrieved from Financial Review: http://www.afr.com/technology

Soergel, A. (2016, June 16). *The Productivity Paradox.* Retrieved from U.S. News and World Report: https://www.usnews.com/news/articles/2016-06-01/productivity-ailing-americas-economic-growth

Stewart, I., De, D., & Cole, A. (2015). *Technology and People: The Great Job-Creating Machine.* London: Deloitte LLP.

The Luddites200 Organising Forum. (2016). *Our heritage, the Luddite Rebellion 1811-1813.* Retrieved from http://www.luddites200.org.uk/theLuddites.html

Turing, A. (1937). On Computable Numbers, with an Application to the Entscheidungsproblem. *Proceedings of the London Mathematical Society, 2.42*, 230-265.

Turing, A. (1950). Computing Machinery and Intelligence. *Mind: A Quarterly Review of Phsychology and Philosophy, 59*, 433-460.

U.S. Department of Energy. (2014). *The Water-Energy Nexus: Challenges and Opportunities.* Washington, D.C.: U.S. Department of Energy.

UK National Archives. (2016). *Luddites.* Retrieved from Power, Politics, and Protest - The Growth of Political Rights in Britain in the 19th Century: http://www.nationalarchives.gov.uk/education/politics/g3/

University of Illinois. (2016, 8). *Machinery Economics.* Retrieved from Farm Analysis Solutions Tools: http://www.farmdoc.illinois.edu/fasttools/spreadsheets/programdescriptions/machineryeconomics.pdf

Veritas Technologies LLC. (2016). *The Databerg Report.* Mountain View, CA: Veritas Technologies LLC.

Wee, R. M. (2004). Soft Errors' Impact on System Reliability. *Electronic Design News*, 69-74.

About the Author

Alex N. Beavers, Jr. is a 40-year veteran high-technology executive, entrepreneur, and author. He currently is a member of the Board of Trustees of the Ringling College of Art and Design, the Advisory Board for Chai Energy, and is the founder of Palma Sola Consulting. Prior to 2017 he was Executive Director of Technology Commercialization at SRI International, founder and CEO of Averatek Corp., founding CEO of Artificial Muscle, Inc., CEO of Thomson Industries, CEO of ITP Systems, CEO of Applicon, Inc., CEO of Schlumberger Automation Systems Asia, General Manager of the General Electric Robotics and Vision Systems Business, and Managing Partner of the PwC High Tech Practice. His academic background includes a BSEE (Vanderbilt University), an MSEE and a PhDEE (University of Houston), and an MBA (Boston University). He has one patent issued. He has authored over 25 papers in technical journals and trade magazines. He has authored one previous book (*Roadmap to the e-Factory*) and was the editor of a second book (*Life Wins!*).

Index

A

abundance 7, 220, 226
actuators 19, 20, 74
agriculture industry...... 121, 123, 205
Alan Turing 23, 24, 88, 235
algorithm.....23, 41, 74, 89, 94, 95, 97, 189
Apple........4, 5, 20, 55, 57, 103, 148, 153, 238
artificial intelligence... 4, 5, 10, 14, 17, 87, 88, 95, 98, 105, 117, 124, 172, 225, 227
automation ... 4, 6, 7, 8, 9, 10, 14, 16, 17, 21, 25, 29, 38, 47, 65, 68, 70, 71, 72, 75, 76, 83, 101, 102, 103, 117, 121, 124, 127, 141, 154, 172, 173, 174, 175, 176, 177, 183, 204, 205, 206, 207, 210, 213, 214, 216, 219, 220, 221, 225

B

Basic Income 215, 216
Battery.......................................26
Big Blue..5
big data 67, 77, 78, 80, 81, 82, 131, 225
Big Data .... vi, 10, 22, 76
Bill Joy..4

C

CAPEX...38, 40, 60, 70, 83, 175, 176

capital expense 38
cloud 4, 27, 50, 77, 192
Collaboration 103
Commercialization...54, 55, 61, 241
Cosmic rays............................194
Cyber Security .... vii, 187

D

David Ricardo 7
decentralization...........82, 83, 84, 145, 230
Decentralization .... vi, 82
Deming 73
digital horizon....... 4, 6, 10, 11, 14, 15, 182, 187, 196, 197, 210, 225, 226, 231
diminishing returns......46, 47, 48, 59, 196, 228
disk drive 18, 24, 54

E

economic transformation......... 57, 58, 62, 63
Edison 108, 109
education industry.........124, 125, 126, 127, 208
Elon Musk......................................4
energy....26, 54, 83, 119, 136, 137, 138, 139, 140, 141, 142, 143, 145, 149, 160, 170, 235
enterprise information 66

F

Facebook 5, 54, 62
Flynn effect 100, 164, 165
Frankenstein 105

G

Google 5, 27, 41, 42, 62, 148, 153, 171, 234
Google, ... 5, 148, 153
government 4, 8, 35, 40, 48, 60, 61, 62, 65, 73, 106, 108, 109, 111, 122, 126, 132, 133, 134, 135, 136, 147, 149, 150, 177, 178, 179, 189, 190, 208, 211, 215, 217, 221

H

Hamburger Automation 172
healthcare industry 93, 127, 129, 132
human intelligence 87, 95, 98, 227

I

IBM 4, 91, 192
income inequality 199, 200, 201, 202
income mobility 200, 202
Innovation 24, 35, 49, 53, 54, 56, 59, 60, 62, 182, 205, 211, 237
Isaac Asimov 105

J

job creation 9, 203, 206
job destruction 183, 204, 206
John Maynard Keynes 8

K

Karl Marx 7

L

Luddite 7, 237, 238
Lyft 150, 233
Lyndon Johnson 8

M

market based economy. 178, 179
Maslow 34, 37, 59, 117, 219, 236
Microsoft 5, 45, 103, 148
Moore's Law 13

N

necessity .... xiv, 13, 54, 228
Nest 19

P

PCB 27, 28
Personalized medicine 81
Peter Diamondis 5, 237
power storage 26
printed circuit board 27
Productivity 9, 33, 35, 36, 37, 42, 46, 48, 49, 59, 62, 65, 68, 121, 122, 123, 137, 138, 151, 152, 182, 228, 236, 237, 238

Q

quality 33, 36, 41, 66, 67, 68, 69, 70, 72, 76, 92, 93, 94, 119, 127, 128, 131, 132, 135, 145, 161, 172, 173, 179, 190, 192, 213, 221, 227, 228, 229, 230
Quality 69

R

R&D 38, 39, 55, 60
return on investment...42, 138, 154, 204, 227
robot....5, 14, 18, 19, 65, 66, 67, 71, 74, 106, 107, 108, 109, 110, 111, 174, 183, 230, 236

S

sensors...... 10, 16, 19, 20, 21, 27, 66, 74, 80, 102, 106, 107, 159, 164, 171, 213, 225, 227, 230
service industries.......72, 76, 128, 151, 214
Sherlock Holmes.................. vi, 97
social services.........136, 211, 212, 213, 216, 217
statistical analysis........ 22, 77, 81
Steve Wozniak............................ 4

T

Technology Adoption Rates...vii, x, 169, 170
Technology deployment 40
Three reasons..... ...226, 228, 230
transportation....33, 60, 62, 76, 84, 102, 136, 138, 145, 147, 149, 150, 153, 171, 183, 206

U

Uber.................... 62, 84, 149, 233

V

Venture Capital........................ 39

W

water.....80, 97, 119, 140, 142, 143, 144, 145
Watson 4, 91, 93
wealth.... 5, 7, 9, 33, 35, 37, 40, 44, 45, 46, 48, 53, 58, 61, 62, 63, 178, 197, 201, 203, 207, 211, 215, 220, 221, 226, 230
Wi-Fi 16, 20, 29
William Lee 6
wireless................. 17, 21, 28, 153
workflow.....68, 70, 72, 73, 75, 76, 134, 228
World Economic Forum...... 5, 237

Y

Yuval Noah Harari.................... 99

www.ingramcontent.com/pod-product-compliance
Lightning Source LLC
Chambersburg PA
CBHW071417180526
45170CB00001B/134